Agroforestry
Systems and Practices

Agroforestry

Systems and Practices

Ramesh Umrani
C.K. Jain

Oxford Book Company
Jaipur ▪ New Delhi

Published By

Oxford Book Company

Regd. Off. "Bony Residency", Gate No. 2
Opp. Tilak Public School, Vishweswariya Nagar,
Gopalpura Road, Jaipur - 302018, India
Ph: 0141 - 2761280, Telefax: 2761381
email - oxfordbookjpr@gmail.com
website - oxfordbookcompany.com

Delhi Office: 102, 1st Floor, Satyam House,
4327/3, Ansari Road, Darya Ganj,
New Delhi - 110002, India
Ph: 011-45652440

Distributed by
KEYA INTERNATIONAL PRESS LLP
E-mail: keyainternationalpressllp@gmail.com

Published in India, 2026

ISBN : 9789380179179

Typeset by Shivangi Computers, Jaipur

PRINTED IN INDIA BY

Global Printing Services,
116, 1st Floor F.I.E. Patparganj Industrial Area, New Delhi-110092.

Contents

Preface

Agroforestry is an integrated approach of using the interactive benefits from combining trees and shrubs with crops and/or livestock. It combines agricultural and forestry technologies to create more diverse, productive, profitable, healthy and sustainable land-use systems. In agroforestry systems, trees or shrubs are intentionally used within agricultural systems, or non-timber forest resources are cultured in forest settings. Knowledge, careful selection of species and good management of trees and crops are needed to optimize the production and positive effects within the system and to minimize negative competitive effects. In some areas, a narrow definition of agroforestry might be defined as simply: trees on farms. Hence, agroforestry, farm forestry and family forestry can be broadly understood as the commitment of farmers, alone or in partnerships, towards the establishment and management of forests on their land. Agroforestry systems can be advantageous over conventional agricultural and forest production methods through increased productivity, economic benefits, social outcomes and the ecological goods and services provided. Biodiversity in agroforestry systems is typically higher than in conventional agricultural systems. Agroforestry incorporates at least several plant species into a given land area and creates a more complex habitat that can support a wider variety of birds, insects, and other animals. Agroforestry also has the potential to help reduce climate change since trees take up and store carbon at a faster rate than crop plants.

The present book describes the state of current knowledge in the rapidly expanding field of agroforestry. Organised into 16 chapters, it reviews the developments in agroforestry and describes the accomplishments in the application of biophysical and socioeconomic sciences to agroforestry. Although the major focus of the book is on the tropics, where the practice and potential of agroforestry are particularly promising, the developments in temperate zone agroforestry are also discussed. The book is designed for students, teachers, and researchers in agroforestry and farming systems.

Ramesh Umrani
C.K. Jain

1

Introduction to Agroforestry

Agroforestry is an integrated approach of using the interactive benefits from combining trees and shrubs with crops and/or livestock. It combines agricultural and forestry technologies to create more diverse, productive, profitable, healthy and sustainable land-use systems. In agroforestry systems, trees or shrubs are intentionally used within agricultural systems, or non-timber forest resources are cultured in forest settings.

Knowledge, careful selection of species and good management of trees and crops are needed to optimize the production and positive effects within the system and to minimize negative competitive effects. In some areas, a narrow definition of agroforestry might be defined as simply. trees on farms. Hence, agroforestry, farm forestry and family forestry can be broadly understood as the commitment of farmers, alone or in partnerships, towards the establishment and management of forests on their land. Where many landholders are involved the result is a diversity of activity that reflects the diversity of aspirations and interests within the community.

Agroforestry systems can be advantageous over conventional agricultural and forest production methods through increased productivity, economic benefits, social outcomes and the ecological goods and services provided. Biodiversity in agroforestry systems is typically higher than in conventional agricultural systems. Agroforestry incorporates at least several plant species into a given land area and creates a more complex habitat that can support a wider variety of birds, insects, and other animals. Agroforestry also has the potential to help reduce climate change since trees take up and store carbon at a faster rate than crop plants.

Agroforestry integrates crops and/or livestock with trees and shrubs. The resulting biological interactions provide multiple benefits, including diversified income sources, increased biological production, better water quality, and improved habitat for both humans and wildlife. Farmers adopt agroforestry practices for two reasons. They want to increase their economic stability and they want to improve the management of natural resources under their care.

A traditional tree farm or nut plantation managed as a single-purpose monocrop is not an agroforestry system. Neither is a woodlot when it's managed for wood products only. Agroforestry involves combining a tree planting with another enterprise such as grazing animals or producing mushrooms or managing a woodlot for a diversity of special forest products. For example, an agroforestry system might produce firewood, biomass feedstocks, pine-straw mulch, fodder for grazing animals, and other traditional forestry products. At the same time, the trees are sheltering livestock from wind or sun, providing wildlife habitat, controlling soil erosion, and in the case of most leguminous species fixing nitrogen to improve soil fertility. Agroforestry practices in use in the United States include alleycropping, silvopasture, windbreaks and shelterbelts, riparian buffer strips, and forest farming (special forest products.)

Among Third World countries, there is a rising tide of concern that the practice of classical forestry—i.e., the raising and management of long-term forestry crops on a massive scale to produce large-dimension timber for commercial or industrial purposes—only increases the disparity in the distribution of wealth. Classical forestry, it is believed, makes a few wealthy individuals richer and the great number of rural poor worse off than ever before.

There is likewise increasing evidence to show that despite the long-term nature of these forestry activities, there is minimal beneficial socio-economic impact upon the rural populations in terms of economic opportunities and community stability; and that very little of the wealth extracted from the forests filters down to and remains in the rural communities. It is no wonder, therefore, that rural inhabitants find very little motivation to support government forestry conservation and restoration programmes.

Their lukewarm attitude toward forestry means that, given a range of prospective uses of upland or hilly land, farmers relegate forestry use to the lowest priority and elevate intensive cropping with annual food crops to the highest. The consequences of this land-use decision upon fragile uplands are well known: general site degradation through soil and nutrient loss, and serious downstream impacts in the form of sedimentation, water pollution, adversely altered hydrologic character, and overall reduction in productivity and sustainability.

Many observers believe that it is both necessary and possible to alter the nature of forestry practice in developing countries in the following ways:

(1) make forestry smaller in scale (village rather than industrial);

(2) make forest resources (land and vegetation) accessible for use by rural populations rather than keeping them as the monopolistic domains of industrial firms; and

(3) integrate forestry more closely with farm operations in order to involve the farmers more intimately and enable their food crops to benefit from the supportive role of inter-cropped trees.

In short, forestry practice should be converted from industrial forestry to village or community forestry.

Besides down-scaling, the other important result of conversion from classical to community forestry would be a widened range of cropping systems. No longer would forestry remain pure forest cropping; it could range from pure forestry at one extreme to the integration of tree crops and annual food crops and livestock at the other. This means the inclusion of agroforestry, a land-use technique that falls under the umbrella of community forestry.

AGROFORESTRY SYSTEMS

Agroforestry has been loosely defined as "trees plus any other crop", or as "combining trees with food crops", but the most objective and comprehensive definition advanced so far is as follows:

> Agroforestry is a system of land use where woody perennials are deliberately used on the same land-management unit as annual agricultural crops and/or animals either sequentially or simultaneously, with the aim of obtaining greater outputs on a sustained basis.

This definition has generated some degree of disagreement. Central to the debate is the nature of the perennial crop component of this system. "Perennial woody crops" could embrace fruit trees as well as timber-bearing or forest trees, and using this general term rather than specifying "forest crops", as implied in the name "agroforestry", has caused foresters to fear that the system may be used as a devious way of converting forests into orchards. Those with the opposite view argue, however, that this cannot happen over all forest lands, that such physical factors as soil quality, topography, elevation and accessibility, acting in concert with socio-economic factors, will serve to limit the conversion of forest lands to other uses.

Classification

Some confusion about what agroforestry really is has arisen because of the profusion of names and terms contrived and used by various researchers in different regions. To be sure, many of these terms are not interchangeable, and the confusion becomes more serious when users, unaware of the slight shades of variation in meaning and coverage, use them indiscriminately. It is necessary to bring some order through classification and appropriate nomenclature.

The first task is to find major systems of agroforestry under which the various types or subsystems could be grouped. The above definition of agroforestry shows that the component crops are arranged either temporally or spatially; thus, these two crop arrangements can serve as principal categories, as follows:

Crop Rotation System (based on temporal arrangement of crops)

As the name implies, this system includes all agroforestry types in which the annual food or cash crops are alternated with tree crops over time (Figure 1). There are in this category two types of crop rotation:

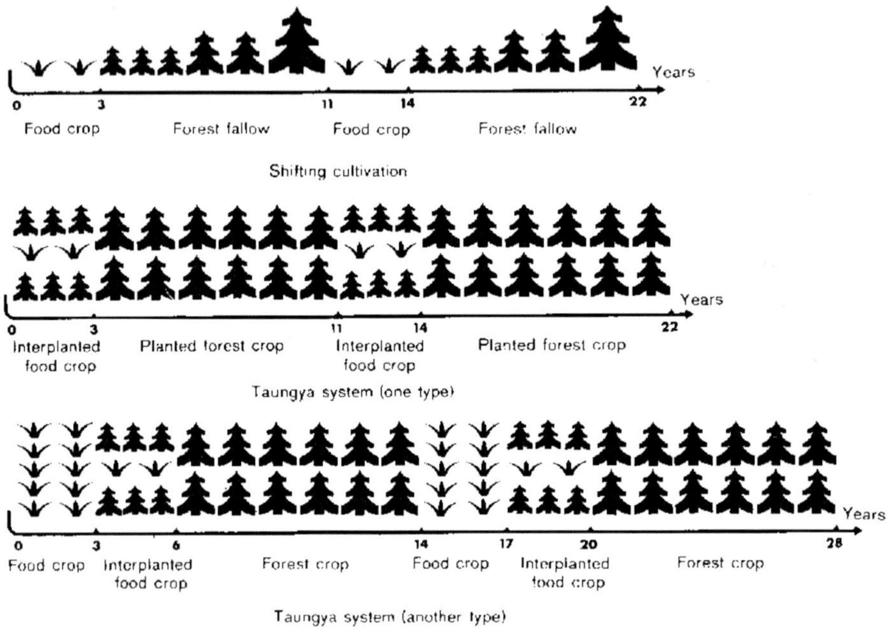

Figure 1: Agroforestry crop rotation system

- Swidden or shifting cultivation is the oldest known agroforestry practice, dating back to the beginnings of the human transition from the hunting and gathering phase to the plant domestication and cultivation system of livelihood. The forest Is cut, dried and burnt to clear the land for planting and to return the nutrients trapped in the forest biomass to the soil to be used by the food crops. Cropping lasts for about two to three years, after which the land is left over a longer period (eight to ten years) to forest fallow in order to rejuvenate the soil and to prepare it for the next cycle of burning and cropping.
- *Taungya,* which had its origins in Burma in the mid-1880s, was the first "modern" (as opposed to "traditional", as in swiddening) agroforestry practice, it was designed as a low-cost governmental approach to the reforestation of open lands. Farmers are temporarily allotted government lands and contracted to plant desired tree species.

While the trees are young and before the canopies close, the farmers are allowed to plant food crops whose yields become theirs entirely and are regarded as their compensation for planting the trees. When food cropping is no longer possible, because of shading, the farmers are transferred to another open area where they repeat the process. Meanwhile, the "abandoned" area is by then a well-established forest and will not be cultivated for annual food crops again until the planted trees reach maturity and are harvested.

The rotation between trees and annuals under the taungya system is slightly different from that in swiddening in the sense that there is a temporal overlap between the crops. Another difference is that the tree crop during the fallow consists of systematically spaced and selected tree species rather than those that are randomly capable of natural regenerating or sprouting in an abandoned swidden.

Intercropping System (based on the spatial arrangement of crops)

Under this system, the annual and perennial crop components are simultaneously present on site but are spaced in such a manner that they became mutually supportive rather than competing. Under such circumstances, they may jointly yield higher outputs per hectare per year. Four subsystems are found under this type:

— *Border tree planting* (Figure 2a) is often found where farmers use lines of trees specifically as boundary markers, live fences, wind-breaks or fire-breaks. These functions complement their services in protecting or stabilizing the site, in producing green manure as organic fertilizer, in producing fodder for farm animals and in producing fuelwood.

Figure 2a: Intercropping system

— *Alternate rows* (Figure 2b) and *alternate strips* (Figure 2c) are referred to in the literature as "alley", "avenue", "corridor", "zonal", or "hedgerow" cropping (a "strip" differs from a row in that it is composed of two or more rows). When positioned across the slopes or along the contours, they are found most effective for erosion control and slope stabilization.

Figure 2b: Intercropping system

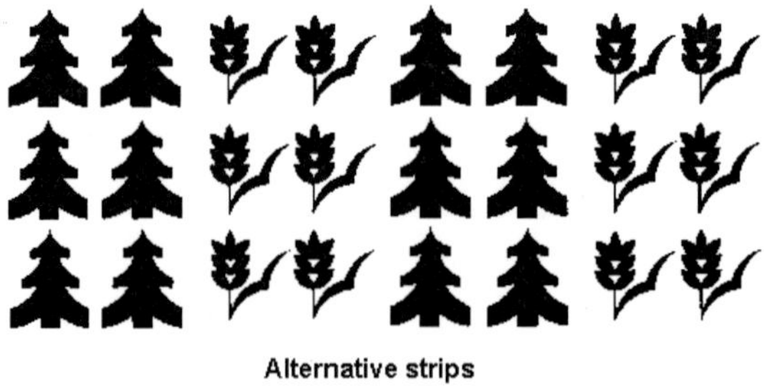

Figure 2c: Intercropping system

— *Random mix* (Figure 2d) is a subsystem that displays no specific or orderly placement of the component crops. While the arrangement appears chaotic, the plants actually occupy their own special ecological niches and are able to coexist very well. Many of the home gardens of Indonesia and the Philippines are of this type.

Figure 2d: Intercropping system

There is still another group of names that need to be classified. The term "agrosilviculture", for instance, is often seen in many publications and has come to be regarded mainly as a synonym for the broad term "agroforestry". However, in reality, it has a different and more precise meaning than similar compound-word terms such as "silvi agriculture", "silvipastoral" and "agrosilvipastoral".

A closer scrutiny of these terms reveals that the hierarchical order of the component crops in the compound names indicates an order of dominance among those crops. For instance, "agrosilviculture" implies that agricultural crops dominate over forest trees, while "silvi-agriculture" indicates that forest crops dominate. This order of dominance is illustrated in a graph (Fig. 3), where cropping systems are shown as a continuum over which appears a range of varying crop combinations.

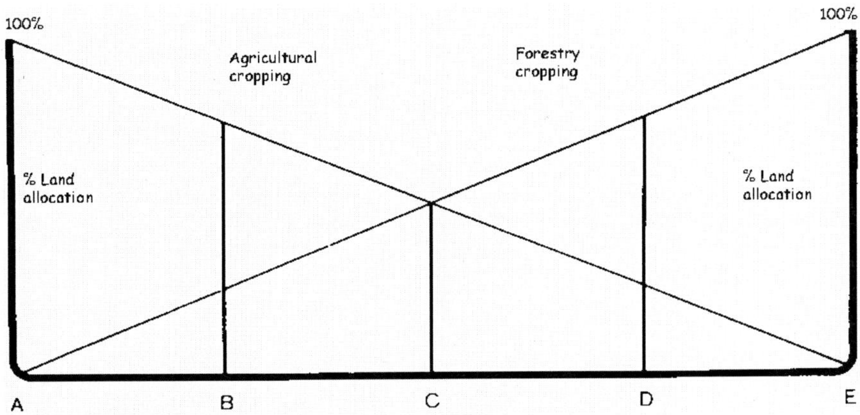

Figure 3: Allocation of crop components in agroforestry

Because of the competition for space between the major crops, forestry and agriculture, an increase in area allocated to one of them automatically results in a reduction of the area allocated to the other—provided, of course, that the entire area is always the sum of the two allocations. In Fig. 3, for instance, point A, at the extreme left, shows that 100 percent of the area is allocated to agriculture. Similarly, on the right (point E), the area is wholly assigned to forestry. At any point between these two extremes will be found agroforestry systems with varying ratios of land allocation. At point B, for example, agriculture dominates, so the system is "agrosilviculture". At point D, on the other hand, forestry is the dominant crop, so the system may be properly termed "silvi-agriculture". A situation can arise, as shown in point C, where the crops get an equal share of the land. In that case, either of I the two terms could be used.

The problem with the use of this graph, which attempts to show the integration of the component crops, is that it is only two-dimensional. It is thus difficult to use in multi-component agroforestry, such as when the livestock element is added. To remedy that weakness, another graphical presentation can be used (Fig. 4). While this new graph is able to show more than two elements or components, it is incapable of showing the degree of dominance of the elements involved. The choice of either Fig. 3 or Fig. 4 as a means for classification would depend on what the user wants to illustrate or emphasize.

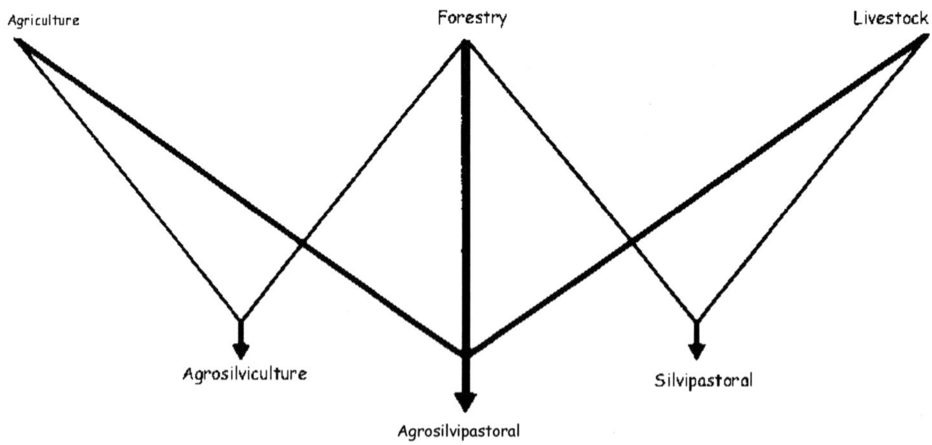

Figure 4: Three types of agroforestry

TROPICAL AGROFORESTRY SYSTEMS

Swiddening

Swiddening as a form of crop-rotation agroforestry is the oldest of the agroforestry systems. In the recent past, when population pressures were low, its practice was tolerated, since the fallow periods between food cropping are sufficiently long to enable the sites to regain their productive capacities, thus making the system sustainable. As pressures mounted, displaced lowland farmers brought and used lowland techniques on hilly lands and drastically shortened or even eliminated forest fallows. The sustainable nature of swiddening was thus lost, with the result that ecological degradation has accelerated both on-site and downstream.

Swiddening has thus become a problem of immense magnitude. Almost all tropical countries are now united in condemning and campaigning against environmentally destructive forms of shifting agriculture. Forest policies and enabling legislations and regulations are directed against the forest farmer. Despite these regulations, however, the farmers persist in practicing swiddening, for want of a more suitable and more acceptable alternative for survival, so the problem has remained.

Taungya

Taungya, which originated in Asia, achieved some degree of success at the start and was adopted by governments of other countries in equatorial Africa and Latin America. Of late, however, farmers have grown wary of the system because it grants no guarantees of tenure over land, and participants usually have to shift from one location to another every three years. Thus, there has been a decline in the number of participant farmers. The irony is that the more successful the cooperating farmers are in establishing the tree crop, the sooner they will be transferred by the government to another open site. Consequently, some of them are motivated to "fail", and they deliberately cause failures by killing young trees.

Another reason for the decline of *taungya* is the growing realization among farmers that their "compensation", solely in terms of the yields of their food crops, is less than those of other reforestation workers under normal wage arrangements. Consequently, government projects now have to supplement the "wage in kind" with cash wages. The only exception to this is in places where population pressures are great: here, *taungya* rights are frequently auctioned.

One important result of conversion from classical to community forestry would be a widened range of cropping systems. Under taungya systems, the irony is that the more successful farmers are, the sooner they are transferred to another site.

Border Tree-planting

On good, level lands, where farmers are not beset by extreme problems of site degradation through erosion and runoff and therefore do not require the protective role of trees, the motivation to intercrop with perennials is low, and the desire to use every square metre of good land for food crops is great. Thus, it is normal to find vast areas of level crop-lands almost treeless.

However, border tree-planting as a form of agroforestry on lowland areas is gaining adherents for several reasons. With the rising prices of energy, including biomass energy, farmers are increasingly producing fuelwood for their own use and for sale instead of being dependent on kerosene or other non-wood fuels. Fast-growing, multipurpose trees are planted along property borders; they are lopped off periodically for fuelwood and their leaves are also harvested and used as fodder or as green manure. In addition, normal litter-fall serves as added green fertilizer for the food crops.

Another frequent reason for border planting is the need for permanent fence posts around the farms or home gardens. Live trees used as posts do not require frequent replacements like the untreated cut poles that decay rapidly under humid tropical conditions. And if prices of wire fences that need to be strung between posts become prohibitive, farmers can plan closely spaced trees that themselves serve as a fence or hedgerow.

Alternate Rows and Alternate Strips

Because most areas occupied by migrant farmers and earmarked for government agroforestry programmes are "problem" areas in the sense that they are on sloping, degradation-prone lands which need either conservation or rehabilitation, the most favoured agroforestry systems have become the alternate-rows or alternate-strips systems, otherwise known as "alley cropping". The reason is that, as indicated earlier, rows or strips of trees planted close together to form contour hedgerows are the most effective vegetative means to stabilize and conserve soils on slopes. This is almost as effective as terracing, and yet it is much easier and less laborious to undertake. Moreover, if the trees selected are capable of fixing nitrogen and are able to regenerate by coppicing, they could serve as a steady source of organic fertilizer to rehabilitate a degraded site.

Random Mix

Random mix remains to this day the mainstay of the traditional smallholder polyculture systems in much of Asia, particularly in the case of the well-known Javanese home gardens and the coconut-based Philippine polyculture. The multi-tiered crowns of the integrated plants, ranging from root crops through grain crops and small fruit trees (cacao, Lansium) to the dominant coconuts, are close replicas of a multi-level tropical rain forest and represent highly efficient utilization of soil surface, vertical space, nutrients, moisture and solar energy. It is probably for this reason that the system has persisted and remained sustainable in these countries over many decades.

On small landholdings that have not been severely degraded, the random mix system could be introduced and maintained, as in the example of the Javanese home gardens, where farmer plots are often just a fraction of a hectare. However, for rehabilitating and maintaining wide areas of badly eroded and depleted lands, random mix may not be as effective as, say, contour hedging or alley cropping.

AGROFORESTRY PRACTICES

Alleycropping

Alleycropping involves growing crops (grains, forages, vegetables, etc.) between trees planted in rows. The spacing between the rows is designed to accommodate the mature size of the trees while leaving room for the planned alley crops. When sun-loving plants like corn or some herbs will be alleycropped, the alleyways need to be wide enough to let in plenty of light even when the trees have matured. Alternatively, the cropping sequence can be planned to change as the trees growth decreases the available light. For example, soybeans or corn could be grown when the trees are very small; then, as the tree canopy closes, forages could be harvested for hay; finally, when the trees are fully grown and the ground is more shaded, grazing livestock or shade-tolerant crops like mushrooms or ornamental ferns could occupy the alleyways. Like all integrated systems, alleycropping requires skillful management and careful planning. Both the crop and the trees have requirements that sometimes necessitate trade-offs between them. The design must allow sufficient room for the equipment needed to service each enterprise. If either crop requires chemical herbicides or insecticides, the other must be tolerant of these treatments. In the case of livestock, there may be periods during and after chemical use when animals must be withdrawn from the area. Livestock can cause damage, even when the trees are fully grown; roots injured by livestock hooves are susceptible to disease. Soil compaction is a danger in wet weather. These examples indicate how crucial planning is to the ultimate success of an agroforestry system.

In most alleycropping systems, trees are planted in straight rows, sometimes with no regard for slope or contour. There are, however, advantages to planting the trees in curves or on the contour. These include the slowing of surface-water movement and the reduction of soil erasion. The trees can be planted in single rows or in blocks of multiple rows between alleys. The first row in a block is planted on the contour line; subsequent rows are planted below the original line according to the slope of the land. The final row of trees in one block is planted parallel to the contour line on which the next block of trees will begin. The width of the tree blocks varies, but the cropping alleyways between them have parallel edges. This design avoids creating point rows within the alleys, thus simplifying crop equipment maneuvers. The width of the alleys is determined by the size of this equipment. If planting on the contour is impractical, another option is to plant trees in curved zigzags so that water running downhill is captured or at least slowed. Islands of trees can offer some of the same advantages if they don't interfere with cropping operations.

In large plantings, fast-growing hardwoods or pines are interplanted as trainers to ensure that the crop trees develop upright, unbranched trunks. Alternatively, the crop trees can be planted close together in the rows, to be thinned and pruned several times as they grow. Although these early-harvested trees may have little market value, their presence during the first years of growth has increased the main crops value. The goal is to produce long, straight sawlogs with few lower branches, for maximum profit at final harvest. Whatever the planting design, trees on the outside edge of a group will grow more side branches, or even a lopsided trunk, resulting in lower-value sawlogs.

Silvopasture

Tree and pasture combinations are called *silvopastoral agroforestry*. Hardwoods (sometimes nut trees) and/or pines are planted in single or multiple rows, and livestock graze between them. Although both the trees and the livestock must be managed for production, some systems emphasize one over the other. Usually, in the early years of establishment, crops or hay are harvested from the planting. Grazing generally begins after two or three years, when the trees are large enough that the livestock can't damage them. In other instances, tree tubes and electric fencing protect the young trees, and grazing begins immediately.

Grazing livestock on silvopasture eliminates some of the costs of tree maintenance. With good grazing management, for example, herbicides and mowing may become unnecessary. Grazing also enhances nutrient cycling and reduces commercial fertiliser costs; the animals remove few nutrients, and their waste is a valuable input for the trees. Well-managed grazing will increase organic matter and improve soil conditions. However, controlling the number of animals per acre, limiting the number of days those animals remain on each site, and avoiding compaction are critical for a successful silvopasture system.

Competition for water between the pasture and the trees may be a concern. In a silvopasture with nut trees, for example, seasonal water shortages during late summer can negatively affect nutfill and the production of fruit buds for next years harvest. Irrigation is justified in such a situation if the trees are being managed for nut production. Water competition may not be as critical for *timber* silvopastures.

Windbreaks or Shelterbelts

Extensive research on windbreaks, also called *shelterbelts,* has been carried out in the U.S. Trees are planted in single or multiple rows along the edge of a field to reduce wind effects on crops or livestock. Windbreaks have been shown to reduce wind impact over a horizontal distance equalling at least ten times the height of the trees. Wind and water erosion are reduced, creating a moist, more favourable microclimate for the crop. In the winter the windbreak traps snow, and any winter crops or livestock are protected from chilling winds. Beneficial insects find permanent habitat in windbreaks, enhancing crop protection.

Although the trees compete for available water along the edges between the windbreak and the crop rows, potentially reducing crop yield near the windbreak, the net effect on productivity is positive. In fact, even on land that's well suited for high-value crops, a windbreak can increase the crop yield of the entire downwind field by as much as 20%, even when the windbreak area is included in the acreage total.

Windbreaks can be designed specifically for sheltering livestock. Studies have shown the economic advantages of providing protection from windchill, a major stress on animals that live outside in the winter. Reduced feed bills, increases in milk production, and improved calving success have resulted from the use of wind-breaks.

Besides providing protection to crops and livestock, windbreaks offer other advantages. They benefit wildlife, especially by serving as continuous corridors along which animals can safely move. Farmers can even develop windbreaks into additional profit centers for the farm hunting leases, selective timber harvests, firewood sales, and special forest products are some of the possibilities.

Any tree species can be used in a windbreak. However, deciduous species, even in multiple rows, will lose effectiveness when they lose their leaves. For year-round use, some of the species selected should be evergreen. Fast-growing trees should be included; it's best to plant deep-rooted, non-competitive species along the edges. Regular deep chisel-plowing along the edges will keep roots from spreading into the crop rows. If some of the trees are harvested periodically, replacements can be planted, establishing a long-term rotation within the windbreak.

Riparian Buffer Strips

Trees, grasses, and/or shrubs planted in areas along streams or rivers are called *riparian buffers* or *filter strips*. These plantings are designed to catch soil, excess nutrients, and chemical pesticides moving over the lands surface before they enter waterways. Such plantings also physically stabilise streambanks. On cropland that is tiled to improve drainage, polluted water can flow directly into streams; constructed wetlands installed in the buffers can capture and clean this drainage water before it enters the stream. Forested areas along streams fulfill other needs of the community at large by storing water and by helping to prevent streambank erosion, which in turn decreases sedimentation downstream. These areas protect and enhance the aquatic environment as well. Shading the water keeps it cooler, an essential condition for many desirable aquatic species. Buffer strips also provide wildlife habitat and can be managed for special forest products.

Crop and livestock farmers, as well as local communities, have become aware of the threat that agricultural practices can pose to pure drinking water. Consequently, there are federal, state, and local government programs to assist in the design and planting of riparian buffer strips. The federal Continuous Conservation Reserve Program can be used for this purpose. The local Farm Services Administration office can advise on this program and other options. Conservation organisations are another potential resource. Some offer conservation easements or trusts when land is permanently withdrawn from agricultural production.

Forest Farming and Special Forest Products

When a natural forested area is managed for both wood products and an additional enterprise, it becomes an agroforestry system. For help with the management of timber, county Extension agents can refer farmers to Extension forestry specialists. These specialists are qualified to give advice on thinning, pruning, and harvesting practices, as well as on marketing options. They may or may not be able to visit the farm for on-site consultation. The Association of Consulting Foresters of America can refer you to private forestry consultants in your area.

Besides producing saw timber and pulpwood, woodlands can generate income from many other products. Established forests offer many non-timber .special forest products that contribute to cash flow without requiring the one-time harvest of old trees. For example, landowners can manage established woods to encourage naturally occurring patches of berries or bitter-sweet. Or they might plant understory crops adapted to the forest type and climate. Growing mushrooms on logs is another, more labor-intensive, possibility; a canopy of either hardwoods or pine will provide the shade needed to maintain moisture for fruiting.

AGROFORESTRY BUSINESS

Establishment Costs and Interim Income

Effort spent at the beginning of an agroforestry project on properly preparing the site and following the recommended planting procedures will pay off well later on.

Depending on the type of project, establishment costs can be considerable. For an alleycropping system or even a windbreak destruction of existing vegetation and deep chiseling or ripping of the soil are minimal requirements. A season of growing a cover crop before planting the trees, and use of mulch or landscape cloth to reduce early competition for water and nutrients, will increase the chances of quick, healthy growth. Lending institutions will likely require a good business plan in order to fund such a project, especially for a beginner. However, government support programs such as the continuous CRP (Conservation Reserve Program) or other program payments will help to defray these costs in some areas of the country. Consult with your local Farm Services Agency about whether such programs would apply to your acreage.

The delay until the income from a new planting begins to pay back these initial costs is a key consideration for most landowners. Alley crops and silvopastures provide income from the area between tree rows in this early stage. In addition, as a stand of same-age trees matures, some trees will be harvested in order to reduce competition as the trees begin to require more space. Although the early thinnings are not likely to be worth very much, the later ones may have some market value. It pays to investigate all the options, including marketing value-added products directly. Hardwood chips could be sold to a land-scaping firm, for instance, or firewood may have nearby customers.

Nut trees produce income from the nuts long before the timber can be harvested. In fact, over the life of the planting, the value of the nut harvest of improved varieties is liable to surpass the value of the wood at final harvest. Black walnut is a valuable timber and nut tree, but it requires a good site and takes a long time (often eighty years) before timber harvest can begin. Early training and pruning, as well as managing fertility and pests, will maximise the value of both crops. Pecans, either native or improved varieties, have some of the same advantages and disadvantages. However, pecan trees are seldom harvested for timber while they are still producing because of the high value of the nut.

In the case of pines, boughs for the ornamental market and pine needles for landscaping mulch provide early income potential. Again, the total value of these products over the life of the stand can be more than that of the timber. The advantage of providing income while trees grow to maturity, however, can be critical to the cash-flow situation of the farm. In every system, the amount and type of management and labor required for interim and final products must be carefully weighed during the design stage.

Marketing

Thorough research into the markets available for each type of tree product is absolutely essential before committing to any forestry enterprise. For most forestry products, the buyer must be relatively close to the site. Otherwise, the transportation costs will eat up potential profits. Although short-rotation woody crops are a relatively new type of forestry without established markets, it is likely that regional markets will develop over time where there are customers such as ethanol producers, electric power producers, and the fiber industry.

Regions where forestry is a longstanding tradition are likely to have markets for all types of wood products (e.g., saw timber, chip and saw, pulpwood). Without such a forestry infrastructure already in place, it is risky to commit to an agroforestry system. However, because private lands are becoming a more important source of tree products, new markets will develop in other regions. It is, of course, difficult to predict where, especially when planning for harvests twenty years or more in the future.

Careful consideration must be given not only to the marketing plan, but to the harvest plan as well. The planting design must accommodate harvest equipment and leave room for maintenance operations. Young trees are easily wounded, and these wounds provide entrance to pest organisms.

Thinning and pruning may generate sales if wisely marketed. This part of the planning process requires the advice of a forestry professional, whether a government agent or a private consultant. Remember that loggers and timber buyers are likely to have their own best interests in mind.

Landowners who want to add value to their forest products have some choices. One way is to certify that the forest and its harvest have been managed according to specified ecological standards. There are currently several eco-label certification programs. Eco-labeling has caught on in Europe where consumer recognition is high, but has not consistently earned premium prices in the U.S. In some cases, landowners can add value themselves, for example by cutting and selling fire-wood. Access to a portable sawmill can enable landowners to saw their own logs into lumber, air dry it, and sell it directly to specialty wood-workers. Other options, like selling pine thinnings as Christmas decorations, require imagination and marketing know-how. Fee hunting or wildlife photography, possibly combined with camping or bed-and-breakfast facilities, might also be considered.

Evaluating Agroforestry Options

Agroforestry systems are much more complex than single-purpose farm or forestry enterprises. Each component of the system the trees as well as the crops or livestock must undergo a series of evaluation procedures: testing against the farm or family goals, evaluating resources, investigating promising options from a longer list of possibilities, making the choice, planning, and then implementing the plan and monitoring progress.

Evaluation of an agroforestry system requires collecting the following information:

- *Farm Accounts. Income* and expenditures for existing enterprises and potential ones, including fixed and variable costs.
- *Planting and Felling Areas. The* program of harvest and planting for each year of the project.
- *Labor and Materials. Includes* the costs of seedlings, fertiliser, herbicides, and insurance, as well as planting, pruning, and thinning expenses.
- *Wood Yields. Predicted* wood-product values by log grade, including cost of harvest and transport.
- *Understory Profiles. Crop* or livestock products, including harvested tree products (nuts, pinestraw), and how production will change through the tree rotation; effects of canopy closure and windbreak benefits.
- *Environmental Impacts. Water* yield, erosion reduction, carbon sequestration, wildlife.
- *Social Effects. Family* and farm goals, support of the rural community, improved visual aesthetics.

Since agroforestry systems in temperate climates have not been studied through several complete rotations, landowners will work with incomplete data during the evaluation process. Yield data from same-age tree plantations must be adjusted for an agroforestry system. Understory competition for water and nutrients, as well as light effects on both understory and tree edges, should be taken into account when projecting yields and expected market values.

Integrating several enterprises necessarily involves multiple interactions. How will each component affect the other for better or worse? How can all operations be managed without damage to other parts of the system? Despite every effort to predict, there will be unforeseen consequences. Advantages and disadvantages will become apparent. It is therefore more critical than usual to continually observe what's happening on the site. If, during planning, certain indicators can be identified as early warning signs, better monitoring will result. An alert manager can avoid losses by quickly noticing problems as they occur.

Agroforestry systems, especially for temperate climates, have not traditionally received much attention from either the agricultural or the forestry research communities. Nevertheless, implementing designs using trees and bushes to enhance crop or livestock production, waste management, and natural resource protection is a step toward a permanent, stable agriculture. Farmers have pioneered many of these systems. Each requires a careful initial design adapted to the site and the farm operation, continuous observation, and a commitment to a long timeline. The resulting farmscape will be beautiful as well as productive, and can be a source of pride for the family and the community.

ENVIRONMENTAL SERVICES OF AGROFORESTRY SYSTEMS

The formal study and promotion of agroforestry systems (AFS), a method of land management used since time immemorial throughout the "old" as well as the "new" worlds, started at the end of the 1970s. Initially the focus was on the description, possible biological and socio-economic advantages as well as disadvantages and the inventory of traditional AFS, mostly in the tropics. This was followed by evaluations of productivity of both existing and novel AFS and more recently studies on the interactions between the component species with a view to improving management and profitability (or reduced risk).

At the end of the 1990s, increased international concern about environmental issues led to new treaties (e.g. Kyoto Protocol) and emphasis on the environmental service functions of alternative land uses. It was rapidly recognized that AFS have many advantages over monocultures in terms of the increasing demand for multifunctional agriculture and that AFS provide important environmental services. Other recognized potentials of AFS include aesthetic values, buffering of protected areas and agroecotourism.

The payment of incentives to farmers whose land use protects natural resources and hence provides a service to the local, national and global community is a new option which could contribute to the financial viability of farms. The title of this congress "Forests, source of life" offers an opportunity to emphasize and review this important new focus in AFS programmes; i.e. the quantification and valuation of service functions of tree-crop and/or tree-animal production systems.

Reducing Soil Erosion and Maintain Soil Fertility

The concepts of soil amelioration by trees in AFS have been reviewed by Young and Buresh and Tian among other authors. Soil improvement in AFS is linked to the growth of nitrogen-fixing trees or deep-rooted trees and shrubs that increase nitrogen availability through biological fixation, recycle plant nutrients from depth (especially in dry zones) and build up soil organic matter.

Formal AFS research initially focused on ways of maintaining soil fertility in annual cropping systems by using leguminous shrub species; e.g. in parkland AFS, in alley cropping and tree-improved fallows. Less research has been carried out on "barrier" AFS (alley cropping along the contour of slopes), though the use of strips of grass and other annual species to trap sediments and nutrients, slow runoff and increase infiltration has been widely promoted by non-governmental organizations in Central America and Asia. Although many of these AFS studies gave promising results on-station or in researcher managed on-farm trials, for productivity and soil fertility parameters, adoption of alley-cropping systems was disappointing because of: high labour and land requirements; in some cases because of the lack of commercial or home use products from the tree/shrub component; and the long time required to show positive changes.

Planted tree fallows are a potential solution to declining soil fertility due to shortened fallow periods in areas where slash-and-burn is still practised. Nitrogen availability, determined by inorganic soil nitrogen or aerobic nitrogen mineralization at 0 to 20 cm depth, and crop yields can be significantly higher after a rotation of nitrogen-fixing trees than after other tree species or grass fallows. Relative to herbaceous fallows greater accumulation of organic material and nutrient storage in biomass, increased root density as well as greater vertical extension of tree roots help maintain nutrient stocks by reducing leaching losses or by taking up nutrients from deep layers.

Szott and Palm reported that, in comparison to leguminous herbaceous fallows, leguminous tree fallows greatly increased the total of phosphorus, potassium, calcium and magnesium stocks in the biomass, litter and exchangeable cations/ available phosphorus (soil; 0-45cm). These authors suggested that fast-growing leguminous trees can accelerate restoration of nitrogen, phosphorus and potassium stocks in the crop layer but may not completely restore calcium and magnesium stocks.

The benefits of perennial crop (e.g. coffee and cacao) shade trees include reduced soil erosion as natural litter fall or pruning residues cover the soil and reduce the impact of raindrops, improve soil structure, increase soil nitrogen content and enhance nutrient retention. Although economic analyses of all the above-mentioned systems are available they do not take into account all the short- and long-term benefits of including the trees, such as improvements or maintenance of soil fertility, nor the possible impact on profitability of service function incentives.

Contribution to Water Quantity

The potential of AFS to help secure water supplies (quantity and quality) is the least studied service function. The trees in AFS influence water cycling by increasing rain and cloud interception (with possible negative and positive effects), transpiration and retention of water in the soil, reducing runoff and increasing infiltration. For example, Bharati et al. reported that infiltration in areas cultivated with maize or soya, or under pastures, was five times less than under riparian strips cultivated with a variety of plant and tree species, suggesting that the latter had a much higher potential to prevent surface runoff reaching water courses. Moreover, trees in AFS can cycle nutrients in a conservative manner preventing their loss through nutrient leaching. Hence AFS can reduce ground water contamination by nitrate and other substances that are harmful to the environment and human health. As a result of less runoff and leaching, micro-watersheds with forest cover or AFS that cover a high percentage of the soil surface produce high quality water.

Reducing Emissions of Greenhouse Gases

Highly productive AFS, including silvopastoral systems, can play an important role in carbon sequestration in soils and in the woody biomass (above and underground). For example, in Latin America, traditional cattle management involves grass monocultures which degrade about five years after establishment, releasing significant amounts of carbon to the atmosphere. Veldkamp estimated that the cumulative net release of CO_2 from low productivity pastures (Axonopus compressus) varied from 31.5 (Humitropept soil) to 60.5 Mg C/ha (Hapludand) in the first 20 years after forest clearing. Well managed silvopastoral systems can improve overall productivity, while sequestering carbon, a potential additional economic benefit for livestock farmers. Total carbon in silvopastoral systems varied between 68-204 t/ha, with most carbon stored in the soil, while annual carbon increments varied between 1.8 to 5.2 t/ha. The amount of carbon fixed in silvopastoral systems is affected by the tree/shrub species, density and spatial distribution of trees, and shade tolerance of herbaceous species. On the slopes of the Ecuadoran Andes, total soil carbon increased from 7.9% under open Setaria sphacelata pasture to 11.4% beneath the canopies of Inga sp. but no differences were observed under Psidium guajava. Soils under Inga contained an additional 20 Mg C/ha in the upper 15 cm compared to open pasture.

Contribution to the Biodiversity Conservation

AFS also can play an important role in the conservation of biodiversity within deforested, fragmented landscapes by providing habitats and resources for plant and animal species, maintaining landscape connectivity (and thereby facilitating movement of animals, seeds and pollen), making the landscape less harsh for forest-dwelling species by reducing the frequency and intensity of fires, potentially decreasing edge effects on remaining forest fragments and providing buffer zones to protected areas. AFS cannot provide the same niches and habitats as the original forests and should never be promoted as a conservation tool at the expense of natural forest conservation. However they do offer an important complementary tool for conservation and should be considered in landscape-wide conservation efforts that both protect remaining forest fragments and promote the maintenance of on-farm tree cover in areas surrounding the protected areas or connecting them, e.g. in the Central American Biological Corridor.

The degree to which AFS can serve conservation efforts depends on a variety of factors, including the design and origin of the AFS (particularly its floristic and structural diversity), its permanency in the landscape, its location relative to remaining natural habitat and the degree of connectivity within the habitat, as well as its management and use, particularly pollarding, use of herbicides or pesticides, harvesting of timber and non-timber products and incorporation of cattle, goats, etc. In general, the more diverse the AFS, the lower its management intensity and the nearer it is to intact habitat, the greater its ability to conserve native plant and animal species. Certain AFS, which closely mimic natural ecosystems (e.g. home gardens, agroforests as well as rustic coffee and cacao AFS), provide a variety of niches and resources that support a high diversity of plant and animals, though usually less than that of intact forest. However, even AFS with low tree densities and low species diversity may help in maintaining biotic connectivity.

Equally important is the attitude of local people towards biodiversity conservation and the perceived resulting benefits (products, services) and losses (e.g. crop damage or raiding, loss of animals), which in turn cause local people to favour or discourage native plants and animals. When hunting intensity is high, populations of game species within AFS are unlikely to be viable regardless of whether there is appropriate habitat available.

While there is a growing literature on biodiversity within AFS, important questions still remain about the long-term viability of animal and plant populations in AFS and what will happen to these populations if the surrounding landscape is increasingly deforested. Most studies to date have monitored or inventoried biodiversity within landscapes that still retain some forest cover, have focused on a few taxa and have been conducted on small spatial and temporal scales. Multi-taxa, multi-scale and long-term studies are needed before the true value of AFS for conservation is known.

Despite these limitations in our current knowledge, there is already sufficient evidence that AFS offer more hope for conservation of plant and animal species than the monoculture crops they usually replace. This finding has led to exciting new initiatives to use AFS as tools for conservation in already deforested and fragmented landscapes. Many of these initiatives include either the direct payment to farmers for biodiversity conservation (e.g. the GEF project led by CATIE; payment for environmental services for AFS in Costa Rica) or the certification of products stemming from these AFS as biodiversity or ecologically friendly.

The service functions provided by AFS, such as soil conservation, carbon capture, water quality and biodiversity conservation are gaining the attention of researchers, planners and politicians, but since these benefits accrue over the medium-long term, are not tangible to farmers and/or the beneficiaries are found beyond the farm boundaries, conservation/adaptation of AFS may be severely limited. The introduction/promotion of trees on farms also has many disadvantages from the farmer's point of view, not least of which is competition with existing crops and pastures. Mechanisms to reward farmers for all of the products and services they can provide are required to encourage the use of AFS.

Although some results are already available on the environmental services of selected AFS in selected sites, more research is clearly needed on the potential trade-offs between the different services involved and on the valuation and financial mechanisms required to directly benefit farmers who provide these services. Since women and children are often the main beneficiaries of the products from traditional AFS (e.g. food supplements, fibres, medicines), possible negative impacts of intensifying the use of a particular tree species (e.g. for timber production) also need to be evaluated. Complex integrative studies, which focus on the possible changes in all the potential products and service functions when the tree component of agricultural systems is increased, as well as on productivity and profitability of AFS, are going to be needed to achieve optimal land use. Without doubt, conceptual, process and other models will have to be used to achieve this goal.

TOWARDS SUSTAINABLE AGROFORESTRY

Both scientific and empirical sources of information have advanced the reasons for the greater stability and higher level of productivity of agroforestry in comparison with monocrops. Using these same scientific bases, modified and improved agroforestry systems can be designed and disseminated among traditional farmers for possible adoption. For instance, random mix home gardens can be modified into contour hedges or alternate rows or alley cropping for better site protection and, eventually, for greater and more steady productivity. Provided that farmers are enabled to participate actively in the conceptualization, planning and development of the agroforestry projects and programmes, they will become more readily inclined to adopt the system and thus move closer toward achievement of ecological stabilization and sustained yields in agroforestry.

REFERENCES

Anderson L.S. & Sinclair, F.L. 1993. "Ecological interactions in agroforestry systems". *Agroforestry Abstracts* 54: 57-91.

De las Salas, G. ed. 1979. *Agroforesty systems in Latin America* (Proceedings). CATIE, Turrialba, Costa Rica, 220 pp.

Nair, P. K. ed. 1989. *Agroforestry systems in the tropics*. Dordrecht, The Netherlands, Kluwer. 664 pp.

Steppler, H.A. & Nair, P.K.R. eds. 1987. *Agroforestry: a decade of development*. Nairobi, ICRAF.

Young, A. 1989. *Agroforestry for soil conservation*. Wallingford, UK, C.A.B. International/ICRAF. 276 pp.

2

Agroforestry Practices in India

Trees and forests were always considered as an integral part of the Indian culture. The ancient scriptures and historical records amply support this. The best of Indian culture was born in the forests. The Aryan civilisation was started in our forests and our Rishis who evolved the Hindu religion, lived in forests in complete harmony with nature.

The ashrams were the centers which harmonised agriculture and pasture with trees, animals and birds. It was widely believed that destruction of forests and cutting of trees created famine conditions, where as planting and maintaining trees were regarded as noble acts. In fact, so much has been written in our ancient literature that individuals on their own agricultural fields were doing planting tree in ancient times. Gradually, during recent periods because of increasing population and huge gap between demand and supply, forests were ruthlessly exploited to meet the increasing demand of fuel, fodder and timber.

To overcome this huge burden upon our existing forests, some alternative steps have to be taken to meet the increasing demand of forest produce i.e. production of such items have to be carried outside the forest areas as well. Hence, in the light of ever increasing demand, concept of multiple use of land with multipurpose tree species has become immensely important. In this context, agro forestry, which is a form of multiple land/ use system, should be adopted and encouraged. The reasons for higher production under agro forestry system include:

- Greater efficiency of tree species for photosynthesis.
- Improved soil structure and fertility with increasing effects on crop yield.
- Reduce losses from soil erosion and more closed cycling of organic matter and nutrients.
- Creating better micro climatic conditions for the growth of agricultural crops.

INDIAN AGROFORESTRY SYSTEMS

Agro forestry is not a new system or concept. The practice is very old, but the term is definitely new. Agro forestry means practice of agriculture and forestry on the same piece of land. Bene et al. defined agro forestry as a sustainable management system for land that increases overall production, combines agricultural crops and animals simultaneously. Nair defines agro forestry as a land use system that integrates trees, crops and animals in a way that is scientifically sound, ecologically desirable, practically feasible and socially acceptable to the farmers.

Another widely used definition given by the International Center for Research in Agro forestry (ICRAF) Nairobi, Kenya, that, "agro forestry is a collective name for all land use systems and practices where woody perennials are deliberately grown on the same land management unit as agricultural crops or animals in some form of spatial arrangement or temporal sequence".

Agroforestry systems in India include trees in farms, community forestry and a variety of local forest management and ethnoforestry practices. A wider definition of agroforestry encompasses a variety of practices, including trees on farm boundaries, trees grown in close association with village rainwater collection ponds, crop-fallow rotations, and a variety of agroforests, silvopastoral systems, and trees within settlements. These systems have been presented as a solution to rising fuelwood prices in India resulting from increase in demand and decrease in supply of fuelwood due to forest degradation.

Overall, India is estimated to have between 14,224 millionand 24,602 milliontrees outside forests, spread over an equivalent area of 17 million ha, supplying 49% of the 201 million tonnes of fuelwood and 48% of the 64 million mof timber consumed annually by the country.

Agroforests as Carbon Sinks

Land-management actions that enhance the uptake of CO or reduce its emissions have the potential to remove a significant amount of CO from the atmosphere if the trees are harvested, accompanied by regeneration of the area, and sequestered carbon is locked through non-destructive (non-CO emitting) use of such wood.

Carbon management through afforestation and reforestation in degraded natural forests is an useful option, but agroforestry is attractive because:

(i) it sequesters carbon in vegetation and possibly in soils depending on the preconversion soil C;

(ii) the more intensive use of land for agricultural production reduces the need for slash-andburn or shifting cultivation, which contributes to deforestation;

(iii) the wood products produced under agroforestry serve as a substitute for similar products unsustainably harvested from the natural forest and

(iv) to the extent that agroforestry increases the income of farmers, it reduces the incentive for further extraction from the natural forest for income augmentation.

Evidence is now emerging that agroforestry systems are promising management practices to increase aboveground and soil C stocks to mitigate greenhouse gas emissions. The C sequestration potential of tropical agroforestry systems in recent studies is estimated between 12 and 228 Mg ha, with a median value of 95 Mg ha.

Therefore, based on the global estimates of the area suitable for the agroforestry (585–1215 × 10ha), 1.1–2.2 Pg C could be stored in the terrestrial ecosystems over the next 50 years.

In India, average sequestration potential in agroforestry has been estimated to be 25 tC per ha over 96 million ha, but there is a considerable variation in different regions depending upon the biomass production. However, compared to degraded systems, agroforestry may hold more carbon. For example, the above-ground biomass accumulation in a Central Himalayan agroforestry system has been found to be 3.9 t hayrcompared to 1.1 t hayrat the degraded forestland.

A major uncertainty, and therefore an issue for future research, is that these estimates are mostly derived through biomass productivity and often do not take into account carbon sequestration in the soil. In order to exploit the mostly unrealized potential of carbon sequestration through agroforestry, in both subsistence and commercial enterprises innovative policies, based on rigorous research results, are required.

Enhancing Soil Fertility and Water Use Efficiency

Ecological intensification of cropping systems in fluctuating environments often depends on reducing the reliance on subsistence cereal production, integration with livestock enterprises, greater crop diversification, and agroforestry systems that provide higher economic value and also foster soil conservation. Maintenance and enhancement of soil fertility is vital for global food security and environmental sustainability.

Although currently India is self-sufficient in terms of food production, for a population expected to rise further, the country will need to enhance both food production as well as tree biomass. The next green revolution and concurrent environmental protection will have to double the food production. Maintaining and enhancing the soil fertility of farmlands to grow foodgrains as well as tree biomass can help meet the demand in future.

Ecologically sound agroforestry systems such as intercropping and mixed arable-livestock systems can increase the sustainability of agricultural production while reducing on-site and off-site consequences and lead to sustainable agriculture. In regions where the green revolution has not been able make a dent due to lack of soil fertility, agroforestry may hold promise. A useful path, complementary to chemical fertilizers, to enhance soil fertility is through agroforestry.

Alternate land-use systems such as agroforestry, agro-horticultural, agro-pastoral and agro-silvipasture are more effective for soil organic matter restoration. Soil fertility can also be regained in shifting cultivation areas with suitable species. For instance, a field experiment to study Nfixation efficiency suggests that planting of stem-cuttings and flooding resulted in greater biological N-fixation, 307 and 209 kg N haby *Sesbania rostrata* and *S. cannabina* respectively. Thus, *S. rostrata* can be used as a green manure by planting the stem-cuttings under flooded conditions.

Through a combination of mulching and water conservation, trees in agroecosystems may directly enhance crop yields of coarse grains. For instance, in the arid region of Haryana, the effect of Prosopis cineraria, Tecomella undulata, Acacia albida and Azadirachta indica on the productivity of *Hordeum vulgare* (barley) was found to be positive. *P. cineraria* enhanced grain yield by 86.0%, T. undulata by 48.8%, A. albida by 57.9% and A. indica by 16.8% over the control. Biological yield was also higher under trees than that in the open area. Soils under different tree canopies were rich in organic carbon content, moisture availability and nutrient status.

Recent studies have found that multiple-use species such as *Bambusa nutans* have the potential to help in soil nutrient binding during restoration of abandoned shifting agricultural lands (jhum fallows) in northeastern India under *B. nutans*. A comparison of jhum cultivation and agroforestry suggests that the latter is an option to address the challenges of slash-and-burn.

A study of nutrient cycling, nutrient use efficiency and nitrogen fixation in *Alnus*–cardamom plantations in the eastern Himalaya found that nutrient standing stock, uptake and return were highest in the 15-yr-old stand. Annual N fixation increased from the five-yr-old stand (52 kg ha) to the 15-yr-old stand (155 kg ha) and then declined with advancing age. Thus, *Alnus*–cardamom plantations performed sustainably up to 15–20 years.

There is robust evidence that agroforestry systems have the potential for improving water use efficiency by reducing the unproductive components of the water balance (run-off, soil evaporation and drainage). Examples from India and elsewhere show that simultaneous agroforestry systems could double rainwater utilization compared to annual cropping systems, mainly due to temporal complementarity and use of run-off in arid monsoon regions. For instance, a combination of crops and trees uses the soil water between the hedgerows more efficiently than the sole cropped trees or crops, as water uptake of the trees reached deeper and started earlier after flood irrigation than that of the *Sorghum* crop, whereas the crop could better utilize topsoil water. Integration of persistent perennial species with traditional agriculture also provides satisfactory drainage control to ameliorate existing outbreaks of salinity. Agroforestry systems can also be useful for utilization of sewage-contaminated wastewater from urban systems.

It must be pointed out that although agroforestry systems may reduce crop yield for a variety of reasons, there may be a trade-off. For instance, studies on traditional agroforestry system in Central Indiafound that the effect of residual nitrogen on the yield of rice crop after removal of 15-yr-old *Acacia nilotica* trees resulted in increase in crop yield (12.5 t ha) that was almost equal to the reduction in crop yield suffered during 15 years of tree growth in the agroforestry system. Yield reductions may also be compensated in the long run by microclimate modification. Even when trees are not removed through total harvest, the species combination may be designed for nutrient release that benefits crops.

Chemical characteristics and decomposition patterns of six multipurpose tree species, viz. Alnus nepalensis, Albizzia lebbek, Boehmeria rugulosa, *Dalbergia sissoo, Ficus glomerata* and *F. roxburghii* in a mixed plantation established on an abandoned agricultural land in a village at 1200 m altitude in Central Himalaya, is a case in point. These species gave the highest rates of N and P release during the rainy season. Thus, *kharif* crops (rainy-season crops) are unlikely to be nutrient-stressed, even if leaf litter is the sole source of nutrients to crops in mixed agroforestry. A diverse multipurpose tree community provides not only diverse products, but may also render stable nutrient cycling.

Biodiversity Conservation

Society needs to craft synergies among sustainable livelihoods, the Kyoto Protocol, the Convention on Biological Diversity, and other international instruments. Genetic diversity of landraces and trees in agroecosystems is particularly of immediate concern as there is a danger of erosion in ethnocultivars as well as knowledge that has generated such diversity.

Using agroforestry systems as carbon sinks, and by designing a suitable emissions trading system, the Kyoto Protocol provides a new source of financial support for protection and management of biological diversity. Continued deforestation is a major challenge for forests and livelihoods. In addition, decreasing biological diversity through species reduction in managed agroforestry systems is also emerging as a challenge.

Although agroforestry may not entirely reduce deforestation, in many cases it acts as an effective buffer to deforestation. Trees in agroecosystems in Rajasthan and Uttaranchal have been found to support threatened cavity-nesting birds, and offer forage and habitat to many species of birds. These systems also act as a refuge to biodiversity after catastrophic events such as fire. Agroforestry also leads to a more diversified and sustainable rural production system than many treeless farming alternatives and provides increased social, economic and environmental benefits for land users at all levels. What constitutes enough biodiversity in agroecosystems depends upon the goal in question and will differ depending on whether the aim is to increase yields to support livelihood improvement or deal with salinity, ground-water levels, soil erosion, leaching of nutrients or weed control.

If we are concerned about conserving important biodiversity, then protected areas are the preferred choice, and biodiversity conservation may not be a primary goal of agroforestry systems. Nevertheless, in some cases agroforestry systems do support as high as 50–80% of biodiversity of comparable natural systems, and also act as buffers to parks and protected areas. Landscape mosaics created by the interplay of rainwater harvesting as an adaptation to climate change and consequent growth of vegetation in agroforestry systemsact as a corridor providing avenues for dispersal and gene flow in wildlife population. An example of buffer is provided by agroforestry around Hyderabad–Secunderabad. Biomass assessment within 100 km radius of twin cities suggests that annual increment of trees and forests in the region approximately equals the estimated annual wood and fuelwood intake of cities and villages. This supply has acted to buffer the pressure on natural forests.

Tree diversity indeed can be large in some Indian village ecosystems. A study in Sirsimakki village of Karnataka by Shastri *et al.*found 952 individuals belonging to 93 species in just 1.7 ha of agroecosystem. An additional 44 species on non-agricultural lands in the village ecosystem that included 'soppina betta', minor forest and reserve forest were found. The overall agroecosystem had more trees (556 trees/ha) and diversity (diversity index 3.5) compared to the non-agro ecosystem that had 354 trees/ha and a species diversity of 3.87. The overall village ecosystem tree density of 418.8 per ha, with 144 species in 2238 individuals in the sampled area of 5.34 ha is a useful resource. Furthermore, home-gardens, with tree species varying between 20 and 40 on each unit and with an average area of 376 m, support in all 93 tree species counted in just 1.7 ha.

Thus, although not a substitute for continuous and intact natural systems, fragments of all sizes and shapes, nonetheless, have conservation relevance. Local farmers who plant trees on their small farms are often surprised later by the number of birds and small mammals that begin to populate the area.

Biological Pest Control

Agroforestry systems create a landscape structure that is important for biological pest control. In small-scale, subsistence agriculture in the tropics, traditional farming practices have evolved that provide a sustainable means of reducing the incidence and damage caused by pests, including nematodes. The biodiversity inherent in multiple cropping and multiple cultivar traditional farming systems increases the available resistance or tolerance to nematodes. In structurally complex landscapes, parasitism is higher and crop damage lower than in simple landscapes with a high percentage of agricultural use.

Poverty Eradsication and Food Security

Agroforestry could contribute to livelihood improvement in India, where people have a long history and accumulated local knowledge. India is particularly notable for ethnoforestry practices and indigenous knowledge systems on tree-growing. In terms of household income, Central Indian upland rice fields provide an illuminating economics. The farms often have an average of 20 *Acacia nilotica* trees per ha, of 1 to 12 years of age. Small farms have more tree density. At a ten-year rotation, these trees provide a variety of products, including fuelwood (30 kg/tree), brushwood for fencing (4 kg/tree), small timber for farm implements and furniture (0.2 m), and non-timber forest products such as gum and seeds. Thus, trees account for nearly 10% of the annual farm income – distributed uniformly throughout the year than in rice monoculture – of smallholder farmers with less than 2 ha farm holding. A combination of *Acacia* and rice traditional agroforestry system has a benefit/cost (B/C) ratio of 1.47 and an internal rate of return (IRR) of 33% at 12% annual discount rate during a ten-year period.

In the northeast Indian State of Meghalaya, guava and Assam lemon-based agrihorticultural agroforestry systems (i.e. farming systems that combine domesticated fruit trees and forest trees) gave 2.96 and 1.98-fold higher net return respectively, in comparison to farmlands without trees. Average net monetary benefit to guava-based agroforestry systems was Rs 20,610/ha (US$ 448.00) and to Assam lemon-based agroforestry systems, Rs 13,787.60/ha (US$ 300.00). Such systems are most useful livelihood improvement strategies in the rainfed agriculture of Meghalaya. Similarly, the net present value for the different agroforestry models on six-year rotation in Haryana varied from Rs 26,626 to Rs 72,705 hayr, whereas the B/C ratio and IAR varied from 2.35 to 3.73 and 94 to 389% respectively. Thus, agroforestry has not only uplifted the socio-economic status of farmers, but also contributed towards the overall development of the region.

There are numerous non-timber forest products collected from wilderness for subsistence and cash income. Often, harvesting is unsustainable because of lack of knowledge about silviculture of species and destructive exploitation strategies driven by market forces. Domestication of such species aimed at commercialization and production of valued products can reduce the pressure on natural ecosystems. Domestication of forest fruit trees and other species grown in agroforestry systems offers significant opportunity for livelihood improvement through nutritional and economic security of the poor in the tropics. The wild edible plants form an important constituent of traditional diet in Sikkim Himalaya, where about 190 species are eaten and almost 47 species are traded in local market. Wild edible fruit species have high carbohydrate contentranging between 32 and 88%. Such fruit trees can be taken up for domestication in agroecosystems on priority action. Trees in agroforestry systems can provide host to globally valued products and thus support livelihoods locally. A study of the 8-yr-old agroforestry intervention in Palamau District, Jharkhand found that the community depended solely on rainfed farming and animal husbandry definitely gains positively by agroforestry interventions. Suitable community plantations of non-timber forest products in tribal areas such as Jharkhand can potentially serve the dual purpose of conserving useful species as well as livelihood improvement of local people. Such programmes in tribal areas have enhanced likelihood of success as communities are dependent on the wild resources for livelihood. In Jharkhand, trees in agroecosystems are particularly valued as host to insects that yield marketable products such as silk, lac productsand honey.

Woodcarving industry is emerging as an important source of income to local artisans worldwide. Promotion of species used in woodcarving industry facilitates long-term locking-up of carbon in carved wood and supports local knowledge. It therefore strengthens livelihoods. For example, Jodhpur, Rajasthan has emerged as a major centre of woodcarving, exporting woodcraft worth Rs 60 million annually, facilitated by traditional knowledge and skill, and growing tourism. Suitable agroforestry programmes may enhance the availability of wood in agroecosystems, thereby improving the ability of developing countries to participate in the growing global economy.

GROWTH OF AGROFORESTRY IN INDIA

All nature–society interactions have trade-offs and agroforestry systems are no exception. Although agroforestry is a useful land-use management option, it requires careful planning and studies on the remaining challenges, such as farm yield decline under agroforestry systems. There may not be an entirely convincing rationale for the argument that agroforestry systems are the answer for livelihood improvement.

Although, over the last twenty-five years of research in India has demonstrated the potential of agroforestry and some practices have been widely adopted, the vast potential is yet to be fully exploited. Research is needed to further refine the key points of agreement and also to fill the crucial knowledge gaps. There is, evidently, a major gap in our understanding of how agroforestry systems contribute to/ fit into rural livelihood improvement. Future research is required to remove many of the uncertainties that remain, and also carefully test the main functions attributed to agroforestry against alternative land-use options in order to know unequivocally as to what extent agroforestry has served these purposes.

Agroforestry practices are strongly dependent on access to land within the community. Households that do not have ownership to lands may not be able to benefit from the agroforestry interventions for livelihood improvement, unless market regimes permit their inclusion through valueaddition services.

Trees in a variety of ethnoforestry and agroforestry systems contribute to food security, rural income generation through diversity of products and services, and can enhance nutrient cycling, improve soil productivity, soil conservation and soil faunal activities. Nonetheless, trees in agroforestry systems can also cause competition with the associated food crops. Agroforestry may, thus, reduce the yield of the agricultural produce in farmlands. For instance, in Haryana, *A. indica* and *P. cineraria* did not produce any significant difference in the wheat yield, while *Dalbergia sissoo* and *Acacia nilotica* gave a reduction in yield. *A. nilotica* had a more prominent effect with a reduction of 40 to 60% wheat yield and *D. sissoo* reduced yield by 4 to 30%, but the reduction effectwas only up to a distance of 3 m. Interestingly, species that did not negatively affect the yield are indigenous trees occurring in traditional agroforestry systems, and they are economically more useful for providing multiple benefits. Selection of such species to enrich agroforestry systems shall be useful for local and national food security.

Not all species desirable for livelihood improvement can be grown without designing an optimum species combination. Many fruit-yielding species that are suitable to tolerate highly alkali soil (pH > 10) become susceptible to waterlogging. The desirability for agroforestry systems due to high potential for livelihood improvement requires special techniques for planting. For example, pomegranate trees are unable to tolerate water stagnation. To avoid mortality due to water stagnation during the monsoon, the raised and sunken bed technique may be necessary for agroforestry practices on highly alkali soil.

Designing a sustainable tree mixture for agroforestry systems is another challenge. In agroforestry, differences in functional group composition do have a larger effect on ecosystem processes than does functional group richness alone. Thus, much time and expense need to be invested in finding species or genetic varieties that combine in more diverse agroecosystems to improve total yield. For instance, a five-year field experiment of tree mixtures for agroforestry system in tropical alfisol of southern India involving mango (*Mangifera indica*), sapota (*Achrus sapota*), eucalyptus (*Eucalyptus tereticornis*), casuarina (*Casuarina equisetifolia*) and leucaena (*Leucaena leucocephala*) found that growth of sapota can be enhanced by 17% when grown in mixture with leucaena.

But a reduction of 12% in the growth of mango may occur when co-planted with casuarina or leucaena. Eucalyptus is incompatible with mango and sapota. Many species suffer from root competition and thus selection of tree species with either low root competitiveness or trees with complementary root interaction is of strategic importance in agroforestry systems.

AGROFORESTRY PRACTICES IN UTTAR PRADESH

After creation of Uttaranchal state in the year 2000, the tree cover in Uttar Pradesh has reduced to only 4.46% where as, the State Forest Policy 1998 envisaged that one third of the total geographical area should come under forest/tree cover. Hence, agro forestry is now the only option to increase the desired tree cover of 33%. In Uttar Pradesh, practices of agro forestry vary considerably according to the agro climatic zones, socioeconomic conditions and site-specific tree species.

The benefits of agro forestry is better understood by the farmers in western region of the state, this may be attributed, to the assured market of agroforestry produce because of flourishing wood based industries in the region. Eucalyptus and Poplar are preferred species in the western region, whereas, Shisham and Teak is preferred species in eastern region. Fruit trees also have considerable share of agro forestry particularly in western part of the state.

Higher yields of crops have been observed in forest-influenced soils than in ordinary soils. In the Tarai area of Uttar Pradesh, Taungya cultivators harvested higher yields of crops such as maize, wheat, pulses etc. without fertiliser.Approximately, 20% higher yields of grains and wood have been reported in agro forestry areas of Haryana and western Uttar Pradesh than from pure agriculture. Experiments conducted at IGFRI, Jhansi indicate that the total yield of fodder is more when fodder grasses are grown with fodder trees than pure fodder grass cultivation. Leucaena leucocephala intercropped with agricultural crops and fodder grasses increase the total yield of food grains, fodder and fuel.

Nitrogen fixing trees grown in the agro forestry systems are capable of fixing about 50 -100 Kg N/ha/year. Experience in Punjab, Haryana, Uttar Pradesh, Gujarat and some parts of the southern states indicate that a tree and agriculture crop production system is more productive.

The total production and value of fuel, fodder and small timber in degraded lands are reported to be many times more than the coarse grains usually produced on them. Sanchez stated that, " appropriate agro forestry systems improve soils physical properties, maintain soil organic matter and promote nutrient cycling". Nitrogen fixing trees are mentioned as one of the most promising component of agro forestry system.

The leaf litter after decomposition forms humus, releases nutrients and improves various soil properties, it also reduces the fertiliser needs. Growing of trees and fodder crops is more economical, particularly on marginal lands. Observations taken in hot arid and semi-arid areas of Rajasthan indicate that marginal lands are incapable of sustaining stable and dynamic cultivation of agricultural crops. Silvipasture consisting of growing trees such as Prosopis, Albizia, Zizyphus and Acacia species may provide many times more returns per unit of land than agriculture under such conditions.

AgroforestryPromotion

The National Agriculture Policy, clearly states, "Agriculture has become a relatively unrewarding profession due to generally unfavorable price regime and low value addition, causing abandoning of farming and increasing migration from rural areas." Hence the Policy stresses, "Farmers will be encouraged to take up farm/agro-forestry for higher income generation by evolving technology, extension and credit support packages and removing constraints to development of agro forestry".

Rural people have been practicing tree planting in their farms and homesteads to meet household requirements of fuel, poles, timber and medicinal plants. With the advent of social forestry, diversification in agriculture was encouraged to generate high income and minimise risks in cropping enterprises. Planning Commission, GOI, 2001 for promoting agro forestry, has recommended the following:

- Rather than having a uniform strategy for the whole country, commercial agro forestry should be adopted in irrigated districts of the country.
- A separate strategy should be developed for rain fed areas for environmental security, sustainable agriculture and food accessibility.
- Suitable species for commercial agro forestry may include Acacia nilotica, Bamboo species, Casuarina equisetifolia, Eucalyptus species, Populus deltoides and Prosopis cineraria for different climatic, edaphic and agricultural conditions.
- Specific institutes have been identified for tree improvement and development of clones of specified species.
- Corporate private sector may be encouraged to take up research and development in tree improvement, development of better clones and micro and macro propagation of quality planting material.
- About 100 NGOs may be identified to carry out clonal propagation of seedlings for distribution to farmers at appropriate price and carrying out extension work. Extension activities should include organising farmers, providing them training in planting techniques, protection measures and other silvicultural operations.
- Technological development to diversify usage of agro forestry species will help to ensure a ready market; for example bamboo is getting rediscovered as a potential raw-material for the development of bamboo composites suitable for use in place of wood and wood composites.
- Bamboo technology mission should be started keeping in view the impending gregarious flowering, followed by mass mortality of bamboo, forest fire famine and insurgency. Circumstances warrant formulation of emergency plans for harvesting and processing of bamboo prior to their flowering.

- As more and more farmers are taking up agro forestry, export-import policies should be modified to encourage agro forestry product marketing.
- A system of market regulation to be put in place, including a mechanism of periodic review in order to protect the interest of both producer and consumer of agro forestry produce.
- A suitable market information system needs to be introduced to inform farmers about major buyers, prevailing prices trends, procedure etc.
- All existing laws executive orders relating to tree felling transport, processing and sale should be amended to facilitate agro forestry.
- Commercial agro forestry may be planned in irrigated districts covering 10 m ha. On annual basis, one million ha should be brought under multipurpose tree species identified by the Task Force. The scheme of NABARD for farm/ agro forestry should be expanded and investment of Rs. 100 crores per year should be ensured.
- It is proposed to cover 18 million ha of rain fed areas on watershed basis under agro forestry for conservation of soil and water and plantation of hardy species such as Eucalyptus, Bamboo and Babul. On annual basis 1.8 million ha is proposed for afforestation under various schemes of Rural Development, NAEB and 'food for work' scheme. An investment of Rs. 2700 crores will be required on yearly basis.
- Major states may establish Agro forestry Cooperative Federation for increasing bargaining powers of farmers in marketing of agro forestry products.
- Wood based industries should continue supply of quality planting material to farmers and ensure suitable buy-back arrangement.

AGROFORESTRY PRACTICES IN SOUTH INDIA

During the last five decades, since independence, much of the efforts were directed towards the physical accessibility of food. In the 21st century, ecological access to food might become the most important challenge raising threats towards the existence of mankind. India is one of the 12 mega-diversity areas in the world, and biodiversity constitutes 5 per cent and it is about 2.4% of global land area. This is not a matter of rejoice as because deforestation is also taking place at a rate of 1.5 million ha every year. As a result, serious genetic erosion is taking place wiping out wide varieties of flora and fauna. In addition, extensive and intensive cultivation of crops also resulted in degradation of biodiversity.

The biodiversity degradation has intern created multidimensional and multifold problems in the system. All these intern affected the socio-economic and ecological balance of the country. This phenomena made the policy maker and implementers to take appropriate corrective measures. Localised efforts as well as nation wide campaign were started in this regard. Agroforestry emerged as the soul of biodiversity conservation.

Agroforestry, essentially a mixed cropping system, implies co-existance of farm and forests which can achieve both natural resources and socio-economic sustainability. Therefore, a research study was undertaken in Karnataka and Kerala, India to know the farmers' perception towards agroforestry, farmers' attitude towards agroforestry and adoption behaviour of agroforestry practices and its impact on socio, economic and ecological sustainability. The study was conducted in twenty randomly selected village of Karnataka and Kerala states, India. From each of the 20 selected villages, 10 farmers practicing agroforestry were selected by adopting random sampling procedure. The variable like perception toward agroforestry was measured by using the scale developed by Latha with little modification. In order to measure the attitude of farmers towards agroforestry, suitable scale was developed by following scientific procedure.

Procedure for measuring the impact of agroforestry on social, economic and ecological conditions was also scientifically developed and standardised for this purpose. The validity and reliability of measures were also established and then pre-tested. Interview schedule was used to collect the data from the selected farmers by personal interview method. Each question was explained to the farmers during the course of interview. The onlookers were avoided to a considerable extent throughout the study. The economics of agro-forestry was worked by measuring the output generated from among the twenty randomly selected farmers.

Table 1. Perception level of farmers towards agroforestry

Sl. No.	*Categories*	*Karnataka State Percentage of farmers (n_1=100)*	*Kerala State Percentage of farmers (n_2=100)*
1.	Poor perception	27	31
2.	Better perception	57	37
3.	Good perception	26	32

With regard to perception of farmers towards agroforestry, the data reveled that the majority of farmers had moderately better perception towards the importance and advantageous of going for agroforestry. This findings can be explained on the basis of the fact that agroforestry itself is relatively new concept for farmers. It may take some more time of better understanding and to acquire more information on the importance of agroforestry. Further, adequate efforts might not have been made by concerned developmental agencies to educate and convince illiterate farmers. The farmers of Kerala possessed medium to higher perception towards agroforestry (Table 1.).

The results revealed that majority of the farmers had 'favorable' attitude towards agroforestry. As majority of the farmers had better level of perception, it is logical that attitude of farmers were also favourable. The farmers of Kerala in general had favourable to more favourable attitude towards homestead agroforestry (Table 2).

Table 2. Attitude of farmers towards agroforestry

Sl. No.	*Categories*	*Karnataka State Percentage of farmers (n_1=100)*	*Kerala State Percentage of farmers (n_2=100)*
1.	Less favourable	23	31
2.	Favourable	43	27
3.	More favourable	34	42

Adoption Behaviour

a) *Nursery practices:* Several important practices like standard size of seedbed (010m X1m), proportion of sand, soil and farmyard manure for seed bed preparation (1:1:1), depth of seed sowing (medium depth), height of the seed bed for producing seedlings in poly bags (30 cm), age of seedlings for transplanting, time of seed sowing, periodicity of watering the seed bed, chemical used to break dormancy, fertiliser application and weeding the seed bed are better adopted by farmers. Reasons for poor adoption of practices like inoculation with biofertilisers, seed treatment and recommended quantity of fertiliser application is due to lack of complete knowledge and cost involved and their non-availability.

b) *Pre-planting practices:* Practices like required size of pit of planting, proportion of soil and farm yard manure to fill the pit were adopted by majority of the farmers since all these practices are simple and were guided by field functionaries while they fail to adopt recommended number of seedlings per unit area (not more then 5% of the total area) in the farmland due to lack of knowledge on recommended spacing.

c) *Planting practices:* Majority of the farmers have adopted important technologies like planting of seedlings, water dripping near beds after planting, root trimming for better establishment of seedlings, fertiliser application and control of pests/diseases. The reason must be that all these practices are simple, easy to practice and most of the activities were carried out under the supervision of the extension personnel. Some of the complex technologies like dosage of recommended fertiliser application, dosage of chemicals to be used for pest and disease management and removing bottom leaves while planting were not adopted due to lack of knowledge and also field personnel failed to provide right guidance. Pit method of planting was followed by majority of farmers as it was found good for better establishment of seedlings. Pot irrigation was practiced after planting.

d) *Silvicultural practices:* Farmers failed to adopt Silvicultural management practice mainly because they are ignorant of all such practices like thinning, pruning, pollarding and the time of conducting such operation.

e) *Harvesting and marketing practice:* Most of the farmers have not adopted the harvesting and marketing technologies for both timber and non timber forest produce. This is due to the fact that farmers were processing a less knowledge on harvesting and marketing technologies and also they resorted to untimely felling of immature trees in order to meet their unexpected financial crisis like marriage, diseases, etc.

When overall adoption level of agroforestry practices were considered 39 per cent of the farmers belonged to high adoption category while rest of them i.e. 25 per cent and 36 percent belonged to medium and low adoption categories respectively.

Impact of Agroforestry

Social Parameters

The social impact of agroforestry practices on farmers are presented in Table 3.

Festivals

The farmers have started celebrating their festivals colourfully due to increase in annual income from agroforestry.

Food habits

The data revealed that there was a variation in food habits before and after adopting of agroforestry. With the adoption of agroforestry, farmers have stopped their traditional profession like hunting, gathering forest produces because they themselves started to produce sufficient food grains.

Communication exposure

Adoption of agroforestry nesseciated them to get in contact with field extension functionaries, radio, newspaper etc to gain more information on agroforestry for better adoption.

Migration

Practicing of agroforestry resulted in increased self-employment opportunities through interventions such as nursery raising, mat weaving, basket making etc. This has resulted in gradual decrease in migration.

Table 3. Social, Economic and Ecological impact of agroforestry on farmers

Parameters	*Mean*		*Standard Deviation*		*'t' value*
	After	*Before*	*After*	*Before*	
A. Social parameters					
Festivals	7.9800	4.930	0.6352	1.1393	29.9383
Food habits	4.3200	3.390	0.5026	0.7089	3.2113
Communication exposure	2.3300	0.840	1.7587	0.7877	13.0535
Migration	2.5200	0.520	1.7025	0.7585	12.6876
Nature of occupation	3.5400	2.340	0.4933	1.0172	14.9248
Total	20.690	12.020	2.486	3.024	19.5769
B. Economic parameters					
Family income	6.4770	2.288	3.4748	1.6708	11.9827
Employment status	3.2100	1.410	4.2457	2.6401	5.0806
Livestock possession	2.5010	1.597	0.8819	1.6517	6.1663
Supplementary income	1.3150	0.628	1.8872	1.3319	4.0605
Farm expenditure	0.2624	0.184	0.5351	1.1191	0.6542
Total	13.755	6.107	6.4090	4.890	13.1512
C. Ecological parameters					
Biomass production	6.6200	5.540	1.1701	1.3059	8.9749
Ground water recharge	10.1720	8.268	3.244	1.9651	8.7975
Dependency on forest	1.0000	2.940	0.8528	7.0278	-6.8655
Pests and disease attack	2.1600	1.380	0.7877	0.8278	11.2652
Climatic changes	2.8400	2.120	0.5074	0.7690	9.5595
Total	22.79	23.960	3.2783	3.714	3.4605

Nature of occupation

Farmers were involved in hunting and gathering of minor forest produce in the nearby forests/estates for their livelihood before the introduction of agroforestry. Now they have stopped these occupations and are concentrating only in farming.

Economic Parameter

The economic impact of agroforestry practices on farmers are presented in Table 3.

a) *Family income:* With the adoption of agroforestry farmers started getting more income by selling the fruits and timber every year. Subsidiary activities like mat weaving, basket making, honey collection, sheep / goat rearing etc. are also taken up as an integral part of agroforestry which also intern contributed to the increased family income.

b) *Employment status:* Farmers who used to migrate to other areas in search of employment were reduced to maximum extent as a result of employment generation due to agroforestry which ensure employment throughout the year.

c) *Livestock possession:* Agroforestry ensured good and cheap fodder, which intern increased the number of livestocks.

d) *Supplementary income:* One of the uniqueness of integrated farming system in general and agroforestry in particular is the promotion of traditional subsidiary occupation. This is due to the availability of raw materials for these activities. As a result farmers started many subsidiary ventures like basket making, mat weaving, bamboo crafts etc. These subsidiary occupations, naturally added to the total family income generated from the farmers.

e) *Farm expenditure:* The expenditure incurred by the farmers on the farming has increased marginally, but not significantly.

Ecological Parameters

The economic impact of agroforestry practices on farmers are presented in Table 3.

a) *Biomass production:* There is significant increase in biomass production viz. fodder, fuel, timber etc. with the adoption of agroforestry. Fodder species like *Glyricidia maculata, Sesbania, Luecaena sp.* (Subabul) and other species like *techtona grandis, acacia, Azdirachta indica* (neem), *Cassia siamea, Melia dubia, Casurina* sp. (Casuarina) were grown in agroforestry. This combination not only meets the fodder needs of the cattle and timber needs for agricultural implements but also add organic matter to the soil.

b) *Groundwater recharge:* The findings explicitly indicated that there is a significant improvement in the ground water availability due to tree based farming interventions. Farm pond was one such major interventions made to harvest excess runoff rainwater in agroforestry plots. These farm ponds were located in the upper/ middle catchment of the land, which enhanced the percolation which intern recharge the groundwater table.

c) *Dependency on natural forest:* Reduction of dependency on the natural forest was observed because agroforestry met the fuel, fodder, fruits, fibre and timber needs of the farmers.

d) *Incidence of pest and diseases:* The incidence of pests like mealy bugs, termites, shoot and stem borer were reduced as they were preyed upon by birds, who have made their nests in trees and some birds were also attracted by the fruits in the agroforestry plots.

e) *Climate changes:* There was significant difference in atmospheric temperature, rainfall pattern and celaphic characters before and after the adoption of agroforestry. Possible reason is that tree species might have acted with regard to the modification of micro climatic parameters. Also the tap-root system of various trees acted as a barrier to soil erosion by holding soil particles tightly.

The overall results of this research study has pointed out that the agroforestry can bring significant social, economic and ecological changes that are desirable for the society and hence agroforestry can achieve effectively social, economic and ecological sustainability.

Crop Diversity

Majority of farmers adopted agriculture cum silviculture cum horticulture system of agroforestry. The Karnataka farmers incorporated forest trees such as *Tectona grandis, Acacia nilotica, Grevillea robusta,* Eucalyptus hybrid, *Acacial auriculiformis, Azadirachta indica, Melia dubia, Cassia siamea, Pngamila pinnata, Causuarina equisetifolia, Denrocalamus strictus, Dalbergia catefolia, Pterocarpus marsupune, Albizia lebbek* and fodder green manure crops like *Sesbania grandiflora, Erythrina indica, Glyricidia maculata, Leucaena leuococephala, Ceiba pentandra* and fruit crops like *Mangifera indica, Anacardium oxidentale, Zoziphus marutiana, Tamarindus indica* and *Emlica officinalis.*

As the nearby forests are declared as reserve forests, farmers started facing acute timber and fuel problems. Therefore majority of the farmers adopted the tree species mainly with a reason to meet their fuel and poles logs demand. Also, majority of the farmers were deprived of irrigation facilities, they opted mixed plantation thinking that such ventures will bring more returns.

The agroforestry are unique in Kerala because they are always around the farmers house and hence it is often called as homestead agroforestry. The major crops associated with in the homestead agroforestry system in Kerala are coconut (100%), Banana (99%) vegetables (98%), Mango (95%), Papaya (90%) Tulasi (85%), Pepper (83%), Glyricidia (83%), Arecanut (79%), Jack (76%), Tamarind (75%), etc. Coconut (*Cocos nucifera*) with synonym 'Kalpavriksha', the tree of heaven is the most dominant plantation crop prevalent in homestead of Kerala. It is popularly known as a benevolent provider of all basic needs.

In addition to coconut, perennial plants such as arecanut, mango, jack, pepper, (trailed on any of the trees), tamarind and annuals like banana, solanaceous vegetables, bhendi, amaranthus, tapioca, colocasia, dioscorea, amorphophalllus, ginger, turmeric etc, are grown. All these crops are grown mixed in the coconut garden without any specific row arrangement. Farmers have taken every effort to utilise the both in horizontal and vertical dimensions. Thus, these homesteads consists of multitude of crops representing a multi-tier canopy configuration which ensures a high level of exploitation of environmental resources.

The canopy architecture and pattern of component interaction ensure better energy harvesting and soil exploitation. Homesteads of Kerala, who have sacred grooves are integral part of life, culture and folklore traditions of culture. Some kind of divinity is attached to these mini genetic pool, which directly promote the conservation and sustainable utilisation of these natural eco-systems. The study also revealed that the range of crop species is low in case of farmers with less area and the range increases with an increase in area. The tendency of agroforestry farmer to incorporate various species in his farm increased as the area increased so as to ensure high level of exploration of environmental resources.

Economics of Agroforestry

It is found that farmers were earning at an average of $ 800 or Rs. 31466.20 / every year from one acre of agroforestry plot which is much profitable than any traditional crop. The farmers were also able to save surplus money in the bank, which is a healthy sign of economic sustainability.

FUTURE OF AGROFORESTRY IN INDIA

Although numerous issues are involved with livelihood improvement, agroforestry systems are one option with multifunctional value. In India and other developing countries, the path to sustainable development could be decentralized planning and implementation of strategies that promote local biomass production in agroforestry systems. Such decentralized systems in India can provide critical inputs for livelihood improvement and sustainable development. Along with mitigating the climate change, agroforestry systems can at least partially meet the energy needs of one billion people in India through bioenergy options, by a prudent use of agricultural residues and biomass generated in agroforestry systems. Biomass energybased supply options can create rural wealth and employment necessary for livelihood improvement and sequester large amount of carbon in a decentralized manner. Such strategy would also ensure ecological, economic and social well-being. Thus, an energy and food self-sufficient taluka (a small administrative unit) can be a new model of rural development in India.

Although agroforestry options for carbon sequestration are attractive, they present critical challenges for carbon and cost accounting due to dispersed nature of farmlands and dependence of people on the multiple benefits from agroforestry. Additionally, important concerns regarding monitoring, verification, leakage and the establishment of credible baselines also need to be addressed.

Another challenge is the incentives that promote treegrowing by rural people. Not everyone is willing to adopt agroforestry. We shall need effective communication strategy to extend innovations among people to adopt and maintain agroforestry to supply fuelwood and other products. The likelihood of adoption depends on the availability of lands, progressive attitude of farmers, supportive village institutions, their wealth status and their perceived risk concerning agricultural production.

In order to use agroforestry systems as an important option for livelihood improvement, climatechange mitigation and sustainable development in India, research, policy and practice will have to progress towards:

(i) effective communication with people in order to enhance agroforestry practices with primacy to multifunctional values;

(ii) maintenance of the traditional agroforestry systems and strategic creation of new systems;

(iii) enhancing the size and diversity of agroforestry systems by selectively growing trees more useful for livelihood improvement;

(iv) designing context-specific silvicultural and farming systems to optimize food production, carbon sequestration, biodiversity conservation;

(v) maintaining a continuous cycle of regeneration-harvest–regeneration as well as locking the wood in non-emitting uses such as woodcarving and durable furniture;

(vi) participatory domestication of useful fruit tree species currently growing in wilderness to provide more options for livelihood improvement, and

(vii) strengthening the markets for nontimber forest products.

REFERENCES

Mann, H.S. & Saxena, S.K., eds. 1980. *Khejri* (Prosopis cineraria) *in the Indian desert its role in agroforestry.* CAZRI Monograph No. 11. Jodhpur, India, CAZRI. 93 pp.

Sur, H.S. 1986. Role of windbreaks and shelterbelts on wind erosion, moisture conservation and crop growth - an Indian experience. *In* D.L. Hintz and J.R. Brandle, eds. Proc. Int. Symp. Windbreak Technol., Lincoln, Nebraska, 23-27 June 1986. USDA *Great Plains Agric. Council Publ.*, No. 117. Lincoln, Nebraska, USDA.

Tewari, D.N. 1995. *Agro forestry for increased, productivity, sustainability and poverty alleviation.* International Book Distributors, Dehradun.

Van Crowder, L and Anderson J. 1996. *Integrating Agricultural Research, Education, and Extension in Developing Countries.*

3

Nutrient Cycling in Agroforestry Systems

There are at least 16 essential chemical elements for plant growth. Carbon (C), hydrogen (H), and oxygen (O), obtained from air and water, frequently are not included as 'nutrient' elements. Nitrogen (N), phosphorus (P), potassium (K), calcium (Ca), magnesium (Mg), sulfur (S), iron (Fe), manganese (Mn), zinc (Zn), copper (Cu), boron (B), molybdenum (Mo), and chlorine (Cl) are obtained from the soil and required by all plants. Sodium (Na), silicon (Si), and nickel (Ni) are essential elements for some plant species and, although not required, have positive or beneficial effects on the growth of other species. Cobalt (Co) is essential for nitrogen fixation by legumes. Additional elements, such as selenium (Se), arsenic (As), and iodine (I) are not required by plants, but can be important in plant nutrition since they are essential nutrients for humans and other animals that consume plants. All essential nutrients are equally important for healthy plant growth, but there are large differences in the amounts required. N, P, and K are primary macronutrients with crop requirements generally in the range of 50-150 lb. per acre. Ca, Mg, and S are secondary macronutrients, required in amounts of 10-50 lb. per acre. Micronutrient requirements are generally less than 1 lb. per acre.

Mineral nutrients are obtained by plants through root uptake from the soil solution. Sources of these soluble nutrients in the soil include:

1) weathering of soil minerals,
2) decomposition of plant residues, animal remains, and soil microorganisms,
3) application of fertilizers and liming materials,
4) application of manures, composts, biosolids (sewage sludge) and other organic amendments,
5) N-fixation by legumes,
6) ground rock powders or dusts including greensand, basalt, and rock phosphate,

7) atmospheric deposition, such as N and S from acid rain or N-fixation by lightning discharges, and

8) deposition of nutrient-rich sediment from erosion and flooding.

Mineral nutrients also can be lost from the soil system and become unavailable for plant uptake. Nutrient losses are not only costly and wasteful, but they can be a source of environmental contamination when they reach lakes, rivers, and groundwater. Nutrient losses occur through:

1) *Runoff:* loss of dissolved nutrients in water moving across the soil surface,

2) *Erosion*: loss of nutrients in or attached to soil particles that are removed from fields by wind or water movement,

3) *Leaching*: loss of dissolved nutrients in water that moves down through the soil to groundwater or out of the field through drain lines,

4) *Gaseous losses to the atmosphere:* primarily losses of different N forms through volatilization and denitrification, and

5) *Crop removal*: plant uptake and removal of nutrients from the field in harvested products.

NUTRIENT POOLS IN THE SOIL

In addition to the variety of inputs and outputs, plant nutrients exist in many different forms or 'nutrient pools' within the soil. These pools range from soluble, readily available forms, to weakly bound forms that are in rapid equilibrium with soluble pools, to strongly bound or precipitated forms that are very insoluble and become available only over long time periods. Nutrients in solution can be taken up immediately by plant roots, but they also move with water and can easily leach below the plant root zone or be lost from farm fields. The ideal fertile soil has high nutrient concentrations in the soil solution when crop growth rates are high, but a large storage capacity to retain nutrients when crop needs are low or there is no growing crop.

Exchangeable cations are a short-term storage pool that can rapidly replenish nutrient ions in the soil solution. Soil organic matter releases nutrients slowly as it decomposes, but is an important supply of N, P, S, and micronutrients. Soil minerals and precipitates vary from fairly soluble types (carbonates, sulfates, chlorides) in equilibrium with the soil solution to rather insoluble forms (feldspars, apatite, mica) that release nutrients through reactions with chemical agents such as organic acids. Adsorbed anions, such as phosphate and iron oxides bound to clay and organic matter surfaces, are held strongly and released very slowly, but can contribute to the long-term supply of plant-available nutrients.

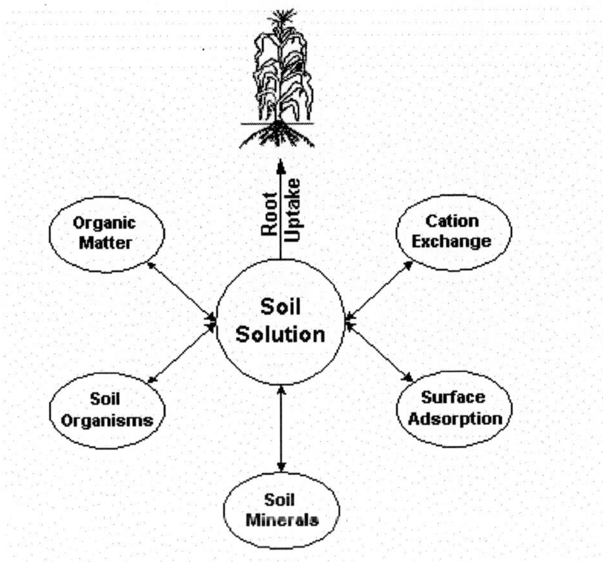

Figure 1. Nutrient pools in the soil

CATION EXCHANGE CAPACITY (CEC)

Clay particles and organic matter have negatively charged sites that hold positively charged ions on their surfaces. CEC protects soluble ions from leaching and loss from the plant root zone. These ions are rapidly exchangeable with other soluble ions, however, so when root uptake depletes the nutrient supply they replenish plant-available cations in the soil solution.

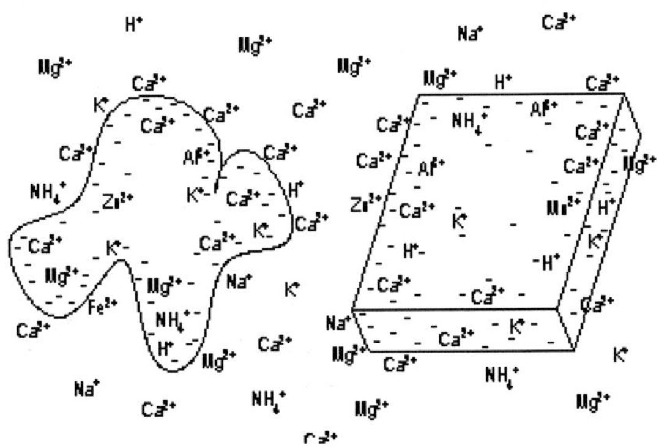

Figure 2. Cation exchange cylce: Clay particles and organic matter

Cation exchange is a major source of nutrients like $K^{+,}$ Ca^{2+}, and Mg^{2+}, as well as NH^{4+} and micronutrient trace metals like Zn^{2+}, Mn^{2+}, and Cu^{2+}.

ORGANIC MATTER

Soil organic matter is a very important factor in soil fertility. It is a reservoir of plant nutrients, has a high CEC, buffers the soil against pH changes, and chelates micronutrients. Organic matter exists in different forms in the soil, ranging from living soil organisms to fresh, readily decomposed plant residues to humus that is very stable and resistant to further degradation. The recycling of plant nutrients through soil organic matter supplies a significant portion of a growing crop's nutrient needs. Stable humus is the organic matter fraction that has a high CEC. Cation exchange helps soils resist changes in pH in addition to retaining plant nutrients. Chelation is the ability of soluble organic compounds to form complexes with micronutrient metals that keep them in solution and available for uptake.

NUTRIENT CYCLES

Soil fertility is maintained when nutrients are efficiently recycled through the soil food web and the soil-plant-animal system. Nutrient cycling is conveniently illustrated indiagrams that can range from very simple to extremely complex.

Basic Plant Nutrient Cycle

The basic nutrient cycle highlights the central role of soil organic matter. Cycling of many plant nutrients, especially N, P, S, and micronutrients, closely follows the Carbon Cycle. Plant residues and manure from animals that are fed forage, grain, and other plant-derived foods are returned to the soil. This organic matter pool of carbon compounds becomes food for bacteria, fungi, and other decomposers. As organic matter is broken down to simpler compounds, plant nutrients are released in available forms for root uptake and the cycle begins again. Plant-available nutrients such as K, Ca, Mg, P, and trace metal micronutrients are also released when soil minerals dissolve.

Nitrogen Cycle

The Nitrogen Cycle is the most complex nutrient cycle. N exists in many forms, both chemical and physical, so transformations between these forms make the N-cycle resemble a maze rather than a simple, circular cycle. Chemical transformations of N, such as nitrification, denitrification, mineralization, and N-fixation are performed by a variety of soil-inhabiting organisms. Physical transformations of N include several forms that are gases which move freely between soil and the atmosphere. Although the N-cycle is very complex, it is probably the most important nutrient cycle to understand. There are two reasons for this:

1) N is usually the most limiting factor for plant growth in terrestrial ecosystems, so there is often a very large crop yield response to additional N, and
2) N in the nitrate form is very soluble and one of the most mobile plant nutrients in the soil, so it can easily be lost from farm fields and become a contaminant in surface waters or groundwater.

Managing N is a critical part of soil fertility management.

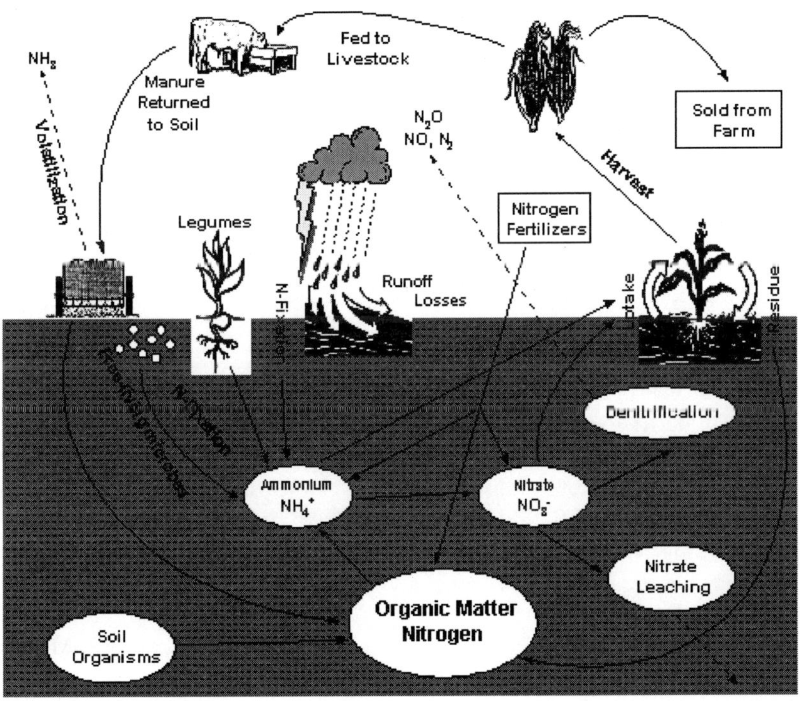

Figure 3. Nitrogen Cycle

Nutrient balance & nutrient budgets

Nutrient cycling is not 100% efficient. There are always some losses or 'leaks' from the cycle, even for natural ecosystems. In farming systems, where products are bought and sold, the balance between nutrient inputs and outputs is easily shifted in one direction or the other. When the balance between inputs and outputs is quantified, a nutrient budget can be calculated. Nutrient budgets can be determined at different scales, from single fields to whole farms to landscapes and even broader regional areas. Strictly speaking, a cycle is a circular, closed-loop pattern, so the nutrient cycles diagramed above are not true cycles. They describe a larger picture where there is movement, or 'flows', of nutrients into and out of smaller systems such as farm fields. Nutrient balances or budgets look at these nutrient flows between different systems.

Whole-Farm Nutrient Budgets

Different types of farms have different patterns of nutrient flow. They vary in patterns of internal movement within the farm as well as in the amounts of external transfers both on-to- and off-of- the farm. Cash-grain and concentrated-livestock farms represent two extremes in nutrient-flow patterns, with mixed crop and livestock farms in an intermediate position. Looking at these three farm types outlines the consequences and challenges faced by a range of different farm types in maintaining soil fertility and using plant nutrients efficiently.

Cash Grain

Cash-grain farms export large amounts of plant nutrients in off-farm grain sales. A 150-bu./acre corn crop, for example, contains about 135-lb. of N, 25-lb. of P, and 30-lb. of K in the grain. When stover or small grain straw is also sold, nutrient losses are even larger, especially for K. To maintain high yields, these nutrients must be replaced. Biologically-fixed N from soybeans in the rotation supplies some N, but large N inputs from forage legumes are not usually a part of systems without livestock to consume the forage. Some soils may have enough residual P, K, or other nutrients to meet crop needs over the short term, but over the long term large amounts of off-farm fertilizer inputs are required to maintain nutrient balance and crop yields in cash-grain systems. In this age of globalization, international grain sales have become an important market for farmers. One consequence of global trade is the associated, world-wide transfer of plant nutrients.

Mixed Crop and Livestock

Farms with both crops and livestock have the potential to recycle a large portion of the nutrients used by crops back to the soil, because about 70% of the nutrients consumed in animal feed are excreted in manure or urine. Efficient recycling depends upon storage, handling, and application methods that minimize losses, and an effective nutrient management plan that applies manure to fields in amounts matching crop needs with the nutrient content of the manure. Within a farm, manure can be a method of transferring nutrients between fields. Depending upon the balance between crop and livestock enterprises, whole-farm nutrient budgets on mixed farms include different amounts of nutrient losses in milk, meat, or eggs, and different levels of nutrient inputs from purchased feed and fertilizer.

Concentrated Livestock

Concentrated animal-feeding operations import large amounts of plant nutrients in purchased grain, forage, and bedding. They are generally net nutrient importers, because purchased inputs exceed nutrient losses from milk, meat, or egg sales. These excess nutrients accumulate in animal wastes that often create disposal or storage problems. High-density livestock operations often have an inadequate land base to efficiently use all the manure they generate, so there is increased risk of water contamination. As livestock operations have become larger, they have also tended to concentrate regionally, resulting in increased geographic separation between feed-grain producers and consumers. Manures are bulky products that are difficult and costly to apply and transport long distances. In many locations it is not economical to recycle the nutrients in animal waste, so long-term storage rather than reuse has become the solution to the waste problem. The net result is increasing transfer of nutrients from one part of the country to another and increased dependence on purchased fertilizer inputs in grain production areas.

NUTRIENT CYCLING IN AGROFORESTS

One of the main tenets of agroforestry is that trees enhance soil fertility, the capacity of the soil to provide essential nutrients for plant growth. There is often confusion between key terms such as nutrient inputs, nutrient outputs, nutrient balances, nutrient cycling and nutrient capital. They all refer to a soil-plant system, usually at the scale of the farmer's field. Nutrient inputs are additions originating from outside the system, such as nitrogen fixed from the air by legumes or the use of chemical fertilizers. Animal manures are inputs if the manure was produced outside the system Nutrient outputs are those that leave the system through crop harvest removals, soil erosion, leaching, gas volatilization and other processes. The nutrient balance is the difference between nutrient inputs and outputs. Nutrient cycling refers to the transfer of nutrients already in the soil plant system from one component to another, for example the release of nitrogen from soil organic matter as ammonium or nitrate and its subsequent uptake by plants.

Other processes involved in nutrient cycling are the return of crop residues such as stover back to the soil; manure and urine deposited by cattle in the system; the incorporation of leguminous green manures into the soil and the transfer of nutrients from trees to crops in agroforestry systems through prunings, leaf drop or root decomposition. The fewer the nutrient losses from the system, the fewer the inputs needed from outside the system to balance the budget. The nutrient capital refers to the soil reserves of nutrients that will be released gradually over a time scale of years or decades. This section examines the role of agroforestry in nutrient cycling in different ecosystems and then focuses on the two main nutrients, nitrogen and phosphorus, in smallholder maize-based systems of Africa.

Agroforestry technologies improve soil conditions via more efficient cycling of nutrients. The potential soil benefits of these technologies depend greatly on local conditions and soil characteristics: tropical Alfisols and Andepts of moderate fertility appear to be particularly suited to agroforestry systems. In one study in western Nigeria, for example, researchers found that planting Leucaena improved the regeneration of bush fallow on an Alfisol. After three years, during which Leucaena was cut annually and left as mulch, the Leucaena fallow resulted in significantly better soil conditions: higher effective cation exchange capacity and exchangeable calcium and potassium levels, compared to the bush fallow.

The existence of nearly closed nutrient cycles between a mature, tropical moist forest and the soil on which it grows has been recognized and studied for about 60 years. Nutrient inputs from atmospheric deposition, biological nitrogen fixation and weathering of primary soil minerals are in balance with nutrient losses owing to leaching, denitrification, runoff and erosion. Nitrogen (N), phosphorus (P), potassium (K), magnesium (Mg), calcium (Ca) as well as sulphur (S) and micro-nutrients are absorbed by forest roots and returned to the soil via the decomposition of litter and roots as well by throughfall and stemflow.

Tropical humid forests accumulate huge quantities of nutrients in their vegetation as the forest grows, with a mature forest reaching steady state values of 700 to 2 000 kg N. 30 to 150 kg P and 400 to 3 000 kg K, Mg or Ca per hectare. The soil also contains large quantities of nutrients. The efficient cycling of nutrients from the soil to the biomass and back to the soil makes possible the lush tropical forest growing on acid, relatively infertile soils of the humid tropics, as long as there are no major biomass removals from the system. Improper harvesting of forests results in major disruptions of this process because large quantities of nutrients are removed from the system and nutrient cycles are disturbed.

Agricultural systems differ from natural systems in one fundamental aspect: there is a net output of nutrients from the site via crop harvest removals. This nutrient removal can result in net negative balances if nutrients are not replaced. Nutrient depletion can be offset by fertilizers, manures from outside the field and other nutrient inputs. This is generally the case in commercial farms of the developed world where such additions, coupled with the recycling of crop residues, have resulted in major nutrient accumulations. Such nutrient buildups, however, sometimes result in the pollution of groundwaters and in algal blooms in streams.

The magnitude of nutrient mining as a result of crop harvests in Africa is huge. Net losses of about 700 kg N. 100 kg P and 450 kg K per hectare during the past 30 years have been estimated for 100 million ha of cultivated land. Smaling's studies on nutrient balances throughout Africa lead the authors to conclude that soil fertility depletion is the fundamental biophysical reason for the declining per caput food production in smallholder farms in Africa. Nutrient mining in Africa, therefore, contrasts sharply with nutrient capital buildups in temperate regions.

Agroforestry, the growing of trees with crops and/or livestock on the same piece of land, is believed to promote a more efficient cycling of nutrients than agriculture. This hypothesis is based partially on studies of the efficient cycling of nutrients from litter to trees in natural ecosystems, and the assumption that trees in agroforestry systems will likewise transfer nutrients to intercropped plants. This is supported by observations of higher crop yields near trees of *Faidherbia albida* in the Sahel and where trees have been recently removed, as in the case of bush and tree fallows. Trees, therefore, can have an effect on soil fertility; however, one must also consider the relative importance of other factors such as soil structure, soil organic matter and the competition for light, water and nutrients before expecting a positive effect on soil fertility in a particular system.

Two key principles set agroforestry systems apart from agricultural or forestry systems: competition and complexity. They in turn determine two desirable properties: profitability and sustainability. All four combine biophysical and socioeconomic issues. The biophysical bottom line of agroforestry is how to manage the competition between the tree component and the crop and/or livestock components for light, water and nutrients to the benefit of the farmer. Although agroforestry systems have been classified in a myriad different ways, there are two functionally different types, simultaneous and sequential systems.

Simultaneous agroforestry is where the tree and the crop components grow at the same time and sufficiently close to each other to allow competition for light, water or nutrients. Examples of this type are alley cropping, contour hedges, parklands, boundary plantings, home gardens and several silvipastoral systems. Simultaneous systems can vary greatly in the relative proportions of trees and crops and in their spatial arrangement. Spatially mixed systems such as coffee and cacao plantations include upper-storey trees that provide nutrients, from litter fall and periodic prunings, and shade to the lower storey tree crops. Spatially zoned systems include hedgerow intercropping where annual crops are grown in alleys between rows of trees that are pruned periodically to provide nutrients to the crops in the alleys.

Competition for nutrients is maximized in simultaneous agroforestry systems, particularity short-duration systems such as alley cropping. The available literature, comprising more than 100 alley cropping experimental data sets ranging from semiarid to humid climates in widely different soils, has recently been analysed. Many factors affect alley cropping performance: the choice of tree and crop species, the alley width, the pruning regime, biomass production, the number of crop cycles, the time and frequency of prunings, tillage, fertilization, weed dynamics, etc. The overall conclusion appears to be clear: the chances that alley cropping will work are limited and very site-specific because, in most cases, the trees' competition for water and nutrients is likely to exceed the fertility benefits from the leguminous mulch additions.

Sequential agroforestry systems are those where the maximum growth rates of the crop and the tree components occur at different times, even though both components may have been planted at the same time and are in close proximity. Examples of this type are shifting cultivation, improved fallows, taungya and some multistrata systems. Competition for growth resources is minimized in sequential agroforestry because the peak demands for light, water and nutrients occur at different times for each component. In tropical ecosystems, both natural and derived, nitrogen and particularly phosphorus frequently limit production. The following discussion focuses on these two most limiting nutrients, using maize as the reference crop.

Nitrogen

Inputs

Trees can provide nitrogen inputs in agroforestry systems by two processes: biological nitrogen fixation (BNF) and deep nutrient capture. Although the magnitude of BNF is methodologically difficult to quantify, overall annual estimates are in the order of 150 kg N per hectare. Empirical evidence, such as the presence of active nodules of leguminous species of the Papilionaceae and Mimosaceae families, indicate that BNF can supply considerable nitrogen inputs to crops via litter in soils that are sufficiently well supplied with phosphorus. This is a definite nutrient input. There is also ample evidence that non-fixing trees, including several species of *Cassia* (recently renamed *Senna*), accumulate as much or more nitrogen in their leaves than nitrogen-fixing legumes, presumably because of their greater root volume and ability to capture nutrients. It is, however, important to note that these non-fixing trees are only cycling nitrogen, not adding inputs to the system.

Deep nutrient capture is the uptake of nutrients by tree roots at depths where crop roots are not present. It can be considered an additional nutrient input in agroforestry systems because such nutrients are leached as far as the crop is concerned. They become an input on being transferred to the soil via tree litter decomposition.

An exciting discovery was made recently in western Kenya regarding subsoil nitrate in nitrogen-deficient nitisols (red alfisols and oxisols with a high iron content), where

Hartemink *et al.* detected nitrate levels in the order of 120 kg N per hectare at subsoil depths of 50 to 200 cm. The authors also found that *Sesbania sesban* fallows depleted this pool, thus capturing a resource that was unavailable to maize. The source of this nitrate pool is believed to be the result of the mineralization of topsoil organic nitrogen, which is relatively high in these soils, followed by the leaching of nitrate from topsoil layers. The nitrate anions are then held in the subsoil by positively charged clay surfaces. What the trees have done, in effect, is to expand the volume of soil used by a considerable amount.

Outputs

A typical maize crop in smallholder African farms yields less than 1 tonne per hectare of grain and requires a plant accumulation of less than 40 kg N per hectare. A 4 tonne per hectare maize crop requires 100 kg N per hectare while a 7 tonne crop requires 200 kg N. Two-thirds of this nitrogen is accumulated in the grain and will be exported on harvest. Much of the remainder, located in the stover, may not be cycled back to the soil because it is frequently fed to livestock outside the system and manure is seldom brought back to the field where the maize was grown. Other losses through soil erosion, leaching and denitrification provide a similar nitrogen output to that of grain harvest removals in the Kisii District of Kenya.

Balance

A negative annual balance of 112 kg N per hectare was calculated by Smaling in the Kisii District, where nitrogen inputs totalled 55 kg and nitrogen outputs 167 kg per hectare. This is considered typical of much of Africa.

Cycling

The release of nitrogen from soil organic matter may contribute most of the 40 kg N per hectare taken up by the average maize crop of 1 tonne per hectare. Considering the variable use of nitrogen fertilizers and the very limited returns of crop residues to the soil, most of the internal cycling in smallholder maize-based systems of Africa is likely to be due to the decomposition of soil organic matter.

Can agroforestry provide for nitrogen requirements? The nitrogen content of 4 tonnes of leguminous leaf mulch dry matter ranges from 60 to 150 kg per hectare. This mulch input rate was also recorded in leguminous fallows in Chipata, Zambia, where maize responds strongly to nitrogen fertilizers, but most farmers can no longer afford their use since the elimination of fertilizer subsidies. Two-year-old *Sesbania sesban* fallows doubled maize yields over a six-year period in comparison with continuous unfertilized maize production. This was accomplished in spite of two years without crop production while the *S. sesban* was growing. A cost-benefit analysis also shows about twice the cumulative net benefit of *S. sesban* fallows over unfertilized maize. *S. sesban* fallows added 128 kg N per hectare to the maize crop, probably mostly from BNF and recycling. The fertilizer recommendation in that area is 112 kg N per hectare, slightly lower than the nitrogen input provided by the *S. sesban* fallow.

Appropriate sequential agroforestry systems such as *S. sesban* fallows appear to be able to replace fertilizer nitrogen applications when maize yields are about 4 tonnes per hectare. At high yield levels comparable with those of commercial farms in the industrialized countries (about 7 tonnes per hectare), organic nitrogen inputs are likely to be insufficient and must be supplemented with inorganic fertilizers. The interaction between organic and inorganic sources of nutrients is essentially a new subject of research in the tropics. Very little is known about it because previous research mainly compared one source against the other.

It appears, therefore, that improved fallow and some alley cropping systems can provide a nitrogen input sufficient to meet low levels of maize yields. The, degree to which this nitrogen is actually taken up by the crop depends on a variety of other factors, including the decomposition rate of organic mulches.

Replenishing Nitrogen Capital

The recovery of leguminous leaf nitrogen incorporated into the soil by the crop (10 to 30 percent) is generally lower than the recovery from nitrogen fertilizers (20 to 50 percent). Organic inputs, however, have an important advantage over inorganic fertilizers in sustainability terms. Much of the remaining 70 to 90 percent of the applied organic nitrogen not utilized by crops is incorporated into active and less active pools of soil organic matter because these mulches also provide a source of carbon. Soil microorganisms need a carbon substrate for growth; they also utilize the nitrogen from organic inputs, resulting in the formation of soil organic nitrogen. Furthermore, part of the nitrogen bound in the more recalcitrant organic inputs will also build up soil organic nitrogen.

Inorganic fertilizers do not contain such carbon sources; therefore, most of the fertilizer nitrogen not used by crops is subject to leaching and denitrification losses, while much of the nitrogen released from organic inputs and not utilized by crops could build soil organic nitrogen capital as well as having an important role in building up soil moisture capacity.

The slow accumulation of soil organic nitrogen with agroforestry or other organic inputs is likely to make a difference in terms of long-term sustainability. This strategy is not new and has been used for centuries in temperate agricultural regions, with crop rotations and winter leguminous cover crops. What is new is the potential to do something similar in the tropics with low input systems that suit smallholder farmers' perspectives. The potential for improved nitrogen management needs to be quantified in agroforestry systems by measuring processes such as mineralization, immobilization, denitrification, volatilization and leaching, along with changes in the soil organic nitrogen pools in systems combining organic and inorganic sources of nitrogen.

Phosphorus

Inputs and Cycling

Agroforestry, however, cannot supply most of the phosphorus required by crops. Leguminous mulches and green manures applied at a realistic rate of 4 tonnes per hectare provide 8 to 12 kg P per hectare. This is about half the phosphorus requirement of a maize crop yielding 4 tonnes of grain per hectare, which accumulates 18 kg P. Therefore, phosphorus is often the critical nutrient in agroforestry and other low external-input systems. Inorganic sources of phosphorus must be applied to agroforestry systems in soils depleted of this element. The strategy is to cycle all the available organic sources first, including manures, and supplement the difference with phosphorus fertilizer inputs. Combinations of organic and inorganic sources of phosphorus may result in a more efficient use of nutrients.

The deep capture of phosphorus is likely to be negligible because of the very low concentrations of available phosphorus in the subsoil. Many agroforestry systems do accumulate phosphorus in their biomass and return it to the soil via litter decomposition, but this is cycling and does not constitute an input from outside the system. However, through cycling, some less available inorganic forms of phosphorus in the soil may be converted into more available organic forms.

Outputs

The two main nutrient loss pathways for phosphorus are harvest removals and soil erosion. While the former is a desirable outcome, the latter is environmentally dangerous since, when eroded, a phosphorus-enriched topsoil can cause eutrification of surface waters. Fortunately, there are well-proven biological erosion control options, such as contour leguminous hedges.

Replenishing Phosphorus Capital

The phosphorus in soils of sub-Saharan Africa is being depleted at a fast rate. Agroforestry cannot be expected to provide additional phosphorus to most farming systems. Therefore, an initiative is being considered by the World Bank and other development agencies to replenish phosphorus through large applications of fertilizers.

It is possible to replenish phosphorus capital in soils with a high phosphorus-fixation capacity (usually identified by their red clayey topsoils). Large applications of rock phosphates or other phosphorus fertilizers could replenish the phosphorus capital of these soils after being fixed and then gradually released by way of desorption from the oxide clay surfaces to plants during the next five to ten years. This is being considered as part of a new approach: investing in natural resource capital. One of the problems with this approach, however, is the need to add acidifying agents to the rock phosphates in order to facilitate their dissolution in most phosphorus depleted African soils, which have pH values of about 6. The decomposition of organic inputs may produce organic acids which can help acidify rock phosphate, and this may be the way to overcome the problem.

MAINTAINING SOIL FERTILITY

Management practices to maximize nutrient cycling & nutrient-use efficiency. Nutrient management is defined as the efficient use of all nutrient sources and the primary challenges in sustaining soil fertility are to:

1) Reduce nutrient losses,
2) Maintain or increase nutrient storage capacity, and
3) Promote the recycling of plant nutrients.

In addition, cultural practices that support the development of healthy, vigorous root systems result in efficient uptake and use of available nutrients. Many management practices help accomplish these goals, including establishing diverse crop rotations, growing cover crops, reducing tillage, managing & maintaining crop residue, handling manure as a valuable nutrient source, composting & using all available wastes, liming to maintain soil pH, applying supplemental fertilizers, and routine soil testing. These beneficial cultural practices have multiple effects on the soil fertility factors described above, which makes it important to integrate their use and examine their effects on the complete soil-crop system rather than just a single component of that system.

Crop Rotations

Growing a variety of crops in sequence has many positive effects. In a diverse rotation, deep-rooted crops alternate with shallower, fibrous-rooted species to bring up nutrients from deeper in the soil. This captures nutrients that might otherwise be lost from the system. Including sod crops in rotation with row crops decreases nutrient losses from runoff and erosion and increases soil organic matter. Growing legumes to fix atmospheric N reduces the need for purchased fertilizer and increases the supply of N stored in soil organic matter for future crops. Biologically-fixed N is used most efficiently in rotations where legumes are followed by crops with high N requirements. Rotating crops also increases soil biodiversity by supplying different residue types and food sources, reduces the buildup and carryover of soil-borne disease organisms, and creates growing conditions for healthy, well-developed crop root systems.

Cover Crops

Growing cover crops can be viewed as an extension of crop rotation and provides many of the same benefits. Growing legume cover crops adds biologically-fixed N. The additional plant diversity with cover crops stimulates a greater variety of soil microorganisms, enhances carbon and nutrient cycling, and promotes root health. The soil surface is covered for a longer period of time during the year, so nutrient losses from runoff and erosion are reduced. This longer period of plant growth substantially increases the capture of solar energy and the amount of plant biomass produced, which in turn increases organic matter additions to the soil. This organic matter is a pool of stored energy in the soil, in addition to a nutrient storage pool, and is the food and energy source for soil organisms. If you look at a farming system as an 'ecosystem', and measure the health or productivity of that ecosystem by its harvest of solar energy, then cover crops increase the health of farming systems by increasing the flow of energy and productive capacity through them.

The extended growth period obtained with cover crops also extends the duration of root activity and the ability of root-exuded compounds to release insoluble soil nutrients. A winter cover crop traps excess soluble nutrients not used by the previous crop, prevents them from leaching, and stores them for release during the next growing season. Cover crops can also suppress weeds which otherwise would compete with crops for nutrients.

Soil and Water Conservation Practices

Soil erosion removes topsoil, which is the richest layer of soil in both organic matter and nutrient value. Implementing soil & water conservation measures that restrict runoff and erosion reduces nutrient losses and sustains soil productivity. Tillage practices and crop residue cover, along with soil topography, structure, and drainage are major factors in soil erosion.

Surface residue reduces erosion by restricting water movement across the soil and tillage practices determine the amount of crop residue left on the surface. Reduced tillage or no-till maximize residue coverage. Water moves rapidly and is more erosive on steep slopes, so reducing tillage, maintaining surface residue, and planting on countour strips across the slope are recommended conservation practices. Rotations and cover crops also reduce erosion.

Soils with stable aggregates are less erosive than those with poor structure and organic matter helps bind soil particles together into aggregates. Tillage breaks down soil aggregates and also increases soil aeration, which accelerates organic matter decomposition. Well-drained soils with rapid water infiltration are less subject to erosion, because water moves rapidly through them and does not build up to the point where it moves across the surface. Drainage improvements on poorly drained soils reduce erosion. Improving drainage also decreases N losses from denitrification, which can be substantial on waterlogged soils, by increasing aeration.

Manure Management

Returning manure to crop fields recycles a large portion of the plant nutrients removed in harvested crops. On farms where livestock are fed large amounts of off-farm purchased feeds, manure applied to crop fields is a substantial source of nutrient inputs to the whole farming system. However, just as nutrients can be lost from the soil, nutrient losses from manure during storage, handling, and application are both economically wasteful and a potential environmental problem.

Soluble nutrients readily leach from manure, especially when it is unprotected from rainfall during storage. Nitrogen is also readily lost through volatilization of ammonia, both during storage and when manure is not incorporated soon after field application. Nutrient losses from manure also occur when it is applied at excessive rates. Analyze manure for its nutrient content and adjust application rates based on crop needs and soil tests. Following heavy manure applications with crops that have high nutrient requirements, especially for N and P, reduces losses and increases nutrient-use efficiency. In addition to nutrient value, manure adds organic matter to the soil and provides benefits such as increased CEC for nutrient retention.

Compost and other soil amendments

In addition to manure, organic amendments such as biosolids, food processing wastes, animal byproducts, yard wastes, seaweed, and many types of composted materials are nutrient sources for farm fields. Biosolids contain most plant nutrients, and are much 'cleaner' than they were twenty years ago, but regulations for farm application must be followed to prevent excessive trace metal accumulation. Composting is a decomposition process similar to the natural organic matter breakdown that occurs in soil.

Composting stabilizes organic wastes and the nutrients they contain, reduces their bulk, and makes transportation and field application of many waste products more feasible. On-farm composting of manure and other wastes also facilitates their handling. Most organic materials can be composted, nearly all organic materials contain plant nutrient elements, and recycling all available wastes through soil-crop systems by either composting or direct field application should be encouraged. These practices build up soil organic matter and provide a long-term, slow-release nutrient source.

Inorganic byproducts also can be recycled through the soil and supply plant nutrients. Available materials vary by region, but rock powder from quarries, gypsum from high-sulfur coal scrubbers, and waste lime from water treatment plants are among the waste products that have been beneficially used. When considering the agricultural use of any byproduct, a thorough chemical analysis and review of possible regulations should be done to avoid soil contamination problems. Even seemingly benign byproducts should be analyzed and field-tested on a trial basis before using them on a large acreage.

Vigorous Root Systems

Vigorous root systems tap nutrient supplies from a larger volume of soil, so management practices that stimulate root growth increase nutrient uptake. Uptake efficiency by extensive, well-distributed root systems results from increases in the amount of root surface area in contact with the soil. The extent of root-soil contact is only about 1-2% of total soil volume, even in the surface 6-inch layer of soil where root density is greatest. For immobile nutrients like phosphorus, root growth to the nutrient is very important. In most soils, phosphate will only move a few millimeters toward a root over the entire growing season.

Root-soil contact is determined by root length, root branching, and root hairs. Root hairs are located just behind the root tip and have a relatively short life span of a few days to a few weeks. Actively growing feeder roots are necessary to continually renew these important locations for nutrient uptake. Nutrient absorbing capacity is also increased by symbiotic associations between soil fungi and plant roots. These fungi, called mycorrhizae, function as an extension of plant root systems. Mycorrhizae obtain food from plant roots and in return increase the nutrient absorbing surface for the plant through their extensive network of fungal strands. Mycorrhizae are particularly important for phosphorus uptake and can increase zinc and copper uptake as well.

Root activity also has direct effects on nutrient availability in the soil. Insoluble nutrients are released and maintained in solution by the action of organic acids and other compounds produced by roots. Nutrients are also released because the soil immediately adjacent to roots, the rhizosphere, often has a lower pH than the bulk soil around it as a consequence of nutrient uptake. The rhizosphere stimulates microbial activity and microbes also release organic acids and other compounds that solubilize nutrients.

A number of soil factors and management practices affect root growth, distribution, and health. Compacted soil layers restrict root penetration, low pH in the subsoil can restrict rooting depth, water saturation and poor aeration inhibit root growth, and roots will not grow into dry zones in the soil. Alleviating these conditions through some of the management practices described above can increase nutrient uptake. Cultural practices that maintain soil biodiversity promote healthy root systems, since an active and diverse microbial population competes with root pathogens and reduces root disease.

Soil Acidity and Liming

Soil pH has strong effects on the availability of most nutrients. This is because pH affects both the chemical forms and solubility of nutrient elements. Trace metals such as Fe, Zn, and Mn are more available at lower pH than most nutrients, whereas Mg and Mo are more available at higher pH than many other nutrients. The ideal soil pH for most crops is slightly acid, about 6.3-6.8, because in that range there is well-balanced availability for all nutrients. This pH range is also optimum for an active and diverse soil microbial population.

Some crops grow better at higher or lower soil pH than 6.3-6.8, usually because of specific nutrient requirements. Blueberries grow best around pH 4.5-4.8 and are Fe deficient when the pH is much over 5. Most crops suffer from Al, Fe, or Mn toxicity when soil pH is that low. Legumes do best at a higher pH than most other crops, due to the high requirement for Mo by N-fixing bacteria.

Limestone is the most commonly used material to increase soil pH. Liming also supplies Ca and dolomitic lime supplies Mg as well. Liming rates depend upon the buffering capacity of a soil in addition to the measured pH. Buffering capacity, or ability to maintain pH within a given range, is related to CEC and increases as clay and/or organic matter content of the soil increases. The lime requirement to raise soil pH the same amount is much larger for fine-textured, high organic matter soils than for coarse-textured, sandier soils. Low soil pH is a more common problem than a pH that is too high, but reducing pH may be necessary for acid-loving plants. Elemental S is the most commonly used material to lower soil pH.

Fertilizer Applications

Many materials can be applied to soil as sources of plant nutrients, but the term 'fertilizer' is usually used to refer to relatively soluble nutrient sources with a high-analysis or concentration. Commercially available fertilizers supply essential elements in a variety of chemical forms, but most are relatively simple inorganic salts. Advantages of commercial fertilizers are their high water solubility, immediate availability to plants, and the accuracy with which specific nutrient amounts can be applied. Because they are relatively homogeneous compounds of fixed and known composition, it is very easy to calculate precise application rates. This is in contrast to organic nutrient sources which have variable composition, variable nutrient availability, and patterns of nutrient release that are greatly affected by temperature, moisture, and other conditions that alter biological activity.

The solubility of commercial fertilizers can also be a problem, because soluble nutrients leach when applied in excess or when large rains occur soon after fertilizer application. Increasing soil cation exchange capacity by increasing organic matter reduces the leaching potential of some nutrients. Management practices that synchronize nutrient availability with crop demand and uptake also minimize leaching. Both application timing and the amount of fertilizer are important. Splitting fertilizer applications into several smaller applications rather than a single, large application is especially important on sandy, well-drained soils. Excess nutrient applications can be eliminated or at least significantly reduced by soil testing on a regular basis, setting realistic yield goals and fertilizing accordingly, accounting for all nutrient sources such as manure, legumes, and other amendments, and using plant tissue analysis as a monitoring tool for the fertilizer program.

Soil Testing

Soil testing is a very important soil management tool and serves as the basis for nutrient recommendations and liming requirements. General guidelines for soil testing include:

1. Soil test each field every 1-3 years, depending upon crop rotation and field history
2. Test for organic matter every 3-5 years to follow trends and make sure levels are increasing or being maintained
3. Have soil tested at a reliable, experienced laboratory.
4. Test at the same time every year, so you can make uniform comparisons over time

Soil Sampling

Collecting a representative soil sample is often the weakest link in a soil testing program. Important guidelines for sample collection include:

1. Divide fields for sampling based on differences in soil type, slope, crop history, tillage, and previous lime, fertilizer, or manure applications
2. If you can think of any reason why one part of a field may be different from another, then sample it separately
3. A single sample should represent no more than 20-25 acres; divide larger fields even if they appear uniform

4. Collect 20-25 soil cores or subsamples from the selected area, mix the soil well, take about a pint of this composite sample and submit it for analysis
5. If the soil is wet, spread in a thin layer and air dry or oven-dry at less than 120 degrees F
6. Alternatively, samples can be collected on a grid system; this requires many more samples, but helps map field variability and is necessary for 'precision' or variable-rate nutrient applications
7. Samples should be taken from the 0-8 inch depth in tilled fields
8. No-till fields require two samples: one from 0-4 inches for pH & lime requirement and a separate 0-8 inch sample for nutrient tests
9. Ridge-till fields are sampled at 0-8 inches, but sample position relative to the ridge (row) is important; either sample between the rows after ridging or one-half the way up the ridge shoulder before ridging
10. Pastures should be sampled at the 0-4 inch depth

Soil Nitrate Testing

Nitrogen is a common limiting factor in crop growth, despite the abundance of N in both the soil and atmosphere. Plant roots grow through soil with about 1,000-lb. of N per acre for every 1% soil organic matter. Plant leaves grow in an atmosphere that is 78% N, so there are about 35,000 tons of N in the column of air above every acre of land. Organic N must be broken down to ammonium and nitrate before it is available to a growing crop. Atmospheric N can be captured by legumes and eventually used by other crops in a rotation, but this also requires cycling through an organic phase.

Organic N from soil organic matter, crop residue, manure, compost, biosolids, and elsewhere is a large source of available N for crops. The use of soil organic N in soil testing is usually unreliable, however, because the rate of N release is variable and unpredictable. It is a biological process that varies with temperature, moisture, aeration, the type of organic compounds being decomposed, and the relative abundance of different types of soil organisms. For this reason, N fertilizer recommendations are commonly based on average requirements for N by the crop being grown, and at a specific, anticipated yield level for that crop. Downward adjustments to this N requirement are made for manure applications, preceding legume crops, soil organic matter content, and other N sources, but these adjustments are based on 'average' conditions rather than resulting from reliable and reproducible chemical tests on the specific soil involved.

Pre-Sidedress Nitrate Test

The majority of the N taken up by most crops is in the nitrate form. Testing for soil nitrate is easily and accurately done, but until recently nitrate testing was only used for fertilizer recommendations in regions with low rainfall and limited leaching. In more humid areas like Ohio, nitrate testing during the normal spring or fall soil-testing periods is not an accurate measure of nitrate availability during the growing season because it is too mobile. The usefulness of soil nitrate testing in many parts of the humid, eastern U.S. has changed, however, with the development of the Pre-Sidedress Nitrate Test (PSNT) for corn. In Iowa, this same soil nitrate test is called the Late-Spring Nitrate Test (LSNT).

Soil samples for the PSNT are collected from the upper foot of soil when corn plants are 6-12 inches tall. This is late enough in the season to measure the effect of spring weather conditions on the amount of organic N released, but still early enough to apply additional N if necessary. In wet springs, nitrate testing at this time also gives information about N losses from pre-plant or starter fertilizer due to leaching or denitrification. The PSNT is a tool for efficient N management that can reduce excessive N applications, especially on manured fields and for corn following legumes. Successful research calibrating the test for Ohio has not been accomplished, although it is used to make N fertilizer recommendations for corn in several nearby states. Research efforts are also underway to extend the use of soil nitrate testing by calibrating it for other crops.

End-of-Season Stalk Test

This is a nitrate test on corn stalks done at the end of the growing season to evaluate the outcome of N-management decisions made that year. Nitrate concentrations in the section of stalk 8-16 inches from the base are used to determine whether N availability was too low, too high, or in the adequate range for high yields. The value of the test is due to the fact that nitrate is removed from the stalk during grain-filling, so residual levels are an indication of N availability during the season. The stalk test is a good example of using plant tissue analysis to give feedback on soil fertility. This type of information permits producers to monitor crop responses on their own farms to different soil test levels and standard fertilizer recommendations. Over time, they can use the accumulated results to adjust these 'average' recommendations to the unique conditions of their farms and soil management practices.

Biological and Chemical Approaches to Nutrient Management

The goals of effective nutrient management are to provide adequate plant nutrients for optimum growth and high-quality harvested products, while at the same time restricting nutrient movement out of the plant root zone and into the off-farm environment. Biological processes in the soil control nutrient cycling and influence many other aspects of soil fertility. Knowledge of these important processes helps farmers make informed management decisions about their crop and livestock systems. How these decisions affect soil biology, especially microbial activity, root growth, and soil organic matter are key factors in efficient nutrient management. Managing soil organic matter and biological nutrient flows is complex because crop residues, manures, composts, and other organic nutrient sources are variable in composition, release nutrients in different ways, and their nutrient cycling is strongly affected by environmental conditions.

Chemical processes in the soil, to a large extent, control mineral solubility, cation exchange, solution pH, and binding to soil particle surfaces. Knowledge of soil chemistry makes it possible to formulate fertilizers that supply readily-available plant nutrients. Management of inorganic nutrient sources is simpler than organic nutrient sources, because of their known and uniform composition and the predictability of their chemical reactions. Chemical and biological processes and their effects on plant nutrients cannot be clearly separated, however, since inorganic nutrients can quickly be incorporated into biological cycles.

REFERENCES

Giller, K.E. &: Wilson, K.F. 1991. *Nitrogen fixation in tropical cropping systems.* Wallingford, UK, CAB I.

Hardy, F. 1936. Some aspects of tropical soils. *Transactions Third International Congress of Soil Science,* 2: 150-163.

Hartemink, A.E., Buresh, R.J., Jama, B. & Janssen, B.H. Soil nitrate and water dynamics in *Sesbania* fallows, weed fallows and maize. *Soil Sci. Soc. Am. J.* (in press).

Kiepe P. & Rao, M.R. 1994. Management of agroforestry for the conservation and utilization of land and water resources. *Outlook Agric.,* 23(1): 17-25.

Kwesiga, F. & Coe, R. 1994. The effect of short-rotation *Sesbania sesban* planted fallows on maize yields. *Forest Ecol. Manage.,* 64: 199-208.

Palm, C.A. 1995. Contribution of agroforestry trees to nutrient requirements of intercropped plants. *Agrofor. Syst.,* 30: 105-124.

Sanchez, P.A. 1976. *Properties and management of soils in the tropics.* New York, Wiley.

Sanchez, P.A., Izac, A.-M.N., Valencia, I. & Pieri, C. 1995. *Soil fertility replenishment in Africa: a concept note.* Nairobi, ICRAF.

4

Silviculture

Concern for the sustainable management of the world's forest resources has never been greater. Across the globe, numerous initiatives are under way to foment the development of forest management plans aimed at maximising the contribution of forests to the achievement of development and socio-economic goals without compromising the present and future condition of the resource base. The discussion on sustainability tends to focus on the need to be environmentally "correct" while ensuring provision of adequate direct or indirect economic benefits to the owners or users of forest lands, thereby justifying their maintenance as forest as opposed to their permanent conversion to another form of land use.

Silviculture, as the element of forestry that deals with the establishment, development, reproduction, care and harvesting of forest vegetation, has the daunting yet essential responsibility of providing the biological and technical options to achieve management objectives. Without appropriate silviculture, sustainable forest management is impossible. As such, it would seem logical for silviculture to be at the forefront of attention, both within and outside the forestry sector. Instead, in some instances silviculture seems to be taken for granted while, in others, it is virtually neglected.

Among some actors, particularly within the forestry sector, there is a tendency to affirm that the technical solutions to the challenges of sustainable forest management already exist and that it is only a matter of applying them appropriately. External actors, on the other hand, especially those with a protectionist perspective, often accuse silviculturists of not having the technical knowledge required to satisfy the new and multiple demands being made on forest resources without endangering the resource. They argue, therefore, that the forests should simply be reserved against further use until research has identified appropriate, widely applicable techniques, except perhaps for use by local people who are believed to have sufficient knowledge of their resource, at least in a micromanagement situation. These same actors often argue for "regreening", but without considering with what species, for what use and under what ecological conditions. This tack can only lead to forests that are without long-term biological or economic viability.

Both of the positions err on the side of extremity and yet both also hold elements of truth. Although tested appropriate techniques do exist for a limited number of forest situations, primarily in the temperate and boreal zones, silviculture in the tropics, particularly in terms of natural forest management, is still in its infancy. On the other hand, the suggestion that insufficient scientific knowledge prevents action is unacceptable; the absence of complete knowledge will not keep forest resources from being used or abused.

SILVICULTURE FOR SUSTAINABLE MANAGEMENT

It used to be generally assumed that tropical moist forest ecosystems perpetually renewed themselves and thus represented an inexhaustible wood potential. However, with the expansion of harvesting, this was quickly found not to be the case. The valuable species that were harvested did not necessarily regenerate and it became apparent that logged-over forests would become impoverished unless silvicultural intervention could maintain or increase their timber potential. Thus the first silvicultural practices were aimed at pragmatic procedures to sustain the wood yields of a few economically desirable species. Techniques varied considerably within and between countries as well as over time, depending on the great diversity and complexity of the forest ecosystem and changing attitudes to natural and artificial regeneration.

With important exceptions, organised silvicultural research did not really mature until after the Second World War, notably in the regions under British and French colonial rule. During the past 50 years, tropical research and silviculture has oscillated between two seemingly contradictory concepts: artificial regeneration (e.g. enrichment and plantations) and natural regeneration and improvement of existing stands.

Initially, forest services undertook to enrich the forest through planting. The main idea was to replace valuable species that had been logged, thereby providing material for future harvesting. Additional benefits were expected from pre-existing trees and natural regeneration. The first operations were conservative and concentrated on species with a demonstrated market value, such as *Tarrietia utilis* (niangon), *Entandrophragma utile* (sipo), *E. cylindricum* (sapelli), Khaya spp. (African mahogany), *Aucoumea klaineana* (Gaboon mahogany) and *Chlorophora excelsa* (iroko).

The first enrichment techniques consisted of minimal intervention planting along narrow alleys cut into forests with wide spacing and had little effect on the environment. This method was intended to provide about 50 first-class trees per hectare and depended on frequent and vigorous intervention. It was used almost everywhere in the tropics and gradually evolved from partial to total clearing of initial stands so as to give full light to dense plantings.

Enrichment methods required ongoing maintenance to give seedlings the best chance of survival. The technical nature of the requirements and the need for careful planning and significant labour inputs were rarely satisfied and the results were not very convincing.

The techniques of natural REGENERATION prevailed for about ten years, in particular between 1950 and 1960. Initial silvicultural attempts in natural forests were conducted at the same time as work on enrichment. There was a multiplicity of methods but the common principle was to boost the regeneration of commercial tree species by one of two approaches.

The first was to improve stands by thinning certain categories of trees, without trying to favour directly natural regeneration because the latter was more easily obtained with more homogeneous, treated stands. This approach was used in Zaire and Gabon.

The second was to promote natural regeneration through cutting, clearing and maintenance to open up the overstorey and allow light to reach the ground to trigger the germination of seeds as well as to stimulate the growth of established seedlings suppressed by the forest canopy. These techniques included the Improvement of Natural Stands and, particularly, the Tropical Shelterwood System developed in Southeast Asia and applied in Africa and America.

These natural regeneration practices were eventually abandoned because of problems with the proliferation of sun-seeking creepers, the growth of vigorous pioneers which blocked the development of valuable species and the difficulty of finding and maintaining the right balance of light. These difficulties were compounded by the many interventions widely spread through time and were difficult to justify technically and economically. This last aspect put a stop to stand improvement. Not only was it impossible to see the results immediately but also the gains in yield obtained were not quantified by substantial prior or accompanying research. By the beginning of the 1960s, the scales tipped in favour of artificial methods.

Indeed, advocates of artificial regeneration could claim higher efficiency and a better use of funds by concentrating field operations over less time and allowing systematic—thus stricter and easier-control. The deliberate choice of the field of action and the species was also a positive factor. The intention was to create a new forest, replacing the existing stand with one of a more even structure by planting one or two dominant species. The focus on dense plantings did not lower the unit cost but did make it possible to concentrate work over time and space and to mechanise much of the work.

As a rule, however, enrichment and natural regeneration as initially conceived did not fulfil the hopes of technicians and managers. Impoverished natural stands could not be naturally regenerated but had to be converted into plantations, and dense plantings were popular. Thus, in the 1960s, single-species plantations on bare ground became the main objective of forest managers and researchers in the tropics. At the same time, there was an increasing mechanisation of clearing and maintenance work while the use of fast-growing species, such as pine or eucalyptus, for the production of pulpwood became popular after an experimental phase.

It is noteworthy that the past ten years have seen a renewed interest in natural forests management. Indeed, plantations have been questioned because of their prohibitive cost for many countries, their lower than expected yields in terms of both quality and quantity and on the grounds of biodiversity conservation. "Artificial" and "natural" techniques should not be seen as conflicting but rather as complementary methods. They must both be used and adapted to a new context; that is, the urgent need to protect and manage the tropical forest ecosystem sustainably. To demonstrate the potentially complementary nature of these techniques, the main results of silvicultural research by the former CTFT and its partners in tropical moist forests, both in natural forests and plantations, are presented below.

PLANTATION SILVICULTURE

The technical and economic difficulties of enrichment work led silviculturists to focus on plantation techniques based on the complete removal of the existing forest. This choice was directly inspired by experience gained with light-demanding species such as *Tectona grandis* (teak), *Aucoumea klaineana* (Gaboon mahogany) and *Terminalia superba* (limba). Most techniques currently recommended by CIRAD-Forêt for establishing and managing artificial stands are based on work done by the CTFT in Gabon and, above all, in Côte d'Ivoire.

Choice of Species

Species suitable for tropical moist forest plantations may be classified into three groups:

— *Species with a slow initial growth,* including *Entandrophragma utile* (sipo), Khaya spp. (African mahogany) and *Tarrietia utilis* (niangon). Today, these are rarely used because of many silvicultural, phytopathological and economic constraints.

— *Species that have a rapid initial growth and are used for industrial biomass production* (e.g. pulp, construction wood, firewood). These include pines, eucalypts and acacias, which can be harvested early (between seven and ten years) and can be planted in degraded forest and savannahs.

— *Species that grow well in monospecific (or mixed) stands destined for timber production.* These species form the main subject of research on the management of tropical moist forests. They include the previously mentioned teak, limba and Gaboon mahogany as well as *Cedrela odorata, Gmelina arborea, Terminalia ivorensis* (framiré) and *Triplochiton scleroxylon* (samba).

Establishment and Management of Stands

The planting of any species, whether exotic or indigenous, must take into consideration specific, limiting ecological factors. Some requirements are specific to each species but a few general silvicultural principles can be stated.

Young seedlings are very quickly threatened by self-propagating, sun-seeking plants. Maintenance must be prompt and regular to stop the canopy closing again. Artificial pruning improves the quality of the final product.

The increasing areas established annually as plantations and the growing shortage of labour have led to a greater use of mechanical site preparation, especially for clearing and maintenance. However, mechanisation requires level sites that are free of tree stumps and debris. Heavy equipment also considerably increases the cost of plantation establishment, but it has proved to be indispensable in large-scale projects, especially where climatic conditions restrict the planting season. It also makes it possible to mechanise maintenance and thus improve the survival and initial growth of young stands.

Planting techniques, from nursery preparation of seedlings to field maintenance during the first years, have been the subject of many trials and experiments. Suitable work programmes, including the scheduling and conduct of operations, have been established for several species. The initial plantation density must be high enough for early canopy closure and depends on the architecture and growth pattern of each species. Suitable stocking numbers may range from 1500 to 2000 stems per hectare for teak, 1100 stems for Gaboon mahogany, *Cedrela* sp. and *Gmelina* sp. and as low as 700 stems per hectare for framiré and limba. High initial stocking enables rapid occupancy of the site, but a thinning schedule must be maintained to achieve maximum growth per unit area.

The extensive investments impose a judicious choice of the vegetative material to be propagated. The genetic improvement of forest species starts with an analysis of genetic variability, choice of the best provenances, phenotypic selection and the creation of seed nurseries for the propagation of quality vegetative material. Genetic improvement can be accelerated through vegetative multiplication by taking cuttings from high-performance clones. This technique has been mastered in *Triplochiton scleroxylon* (samba) and *Gmelina arborea*.

Table 1. Silvicultural regime and expected yields for three plantation species

Characteristic	*Aucoumea klaineana*	*Tectona grandis*	*Terminalia ivorensis*
Final density (*stems/ha*)	120	100	70
Diameter of logs (*cm*)	60	50	50-60
Rotation age (*years*)	44	50	34
Basal area (*m²/ha*)	34	20	20
Bole volume (*m³/ha*)	350	275	250
Productivity (*m³/ha/year*)	8	5.5	7.5

Bole volume production depends on stand characteristics that can be controlled by the forest manager. The choice of cutting diameter has a strong influence on the nature and yield of forest stands and should be determined by technical, silvicultural and financial constraints. Present knowledge allows forecasts of production under prescribed silvicultural regimes, average fertility and well-managed plantations (Table 1).

Role and Suitability of Plantations

To conclude the subject of artificial regeneration, it can be stated that suitable plantation establishment techniques have been documented for many species. Enrichment methods provided useful results, both as underplanting and planting on bare ground, but socio-economic and technical constraints have pushed them aside in favour of intensive mechanised methods. These were justified by increasingly impoverished natural forests, by the technical possibilities of large-scale activities and by the insufficient knowledge of silviculture and natural forest management. The question is whether reforestation is a financially viable option for indebted countries with limited resources. The high cost of plantations and their low profitability are disadvantages, but they only apply in the case of purely speculative economic options. The forest, its management, its protection and its reconstitution by plantation are long-term concepts which are financially unappealing, at least given the current market prices of final products. The only way to make the forest financially viable is to integrate market-driven price increases with unquantified sociological and economic advantages: creation of employment, conservation of biodiversity, protection of the environment and the climate (water production, buffering CO_2 levels, etc.). Moreover, because of the extent of irresponsible deforestation in tropical areas, a plantation often constitutes the only forest alternative, even if the new forest forms a simplified ecosystem.

SILVICULTURE IN NATURAL FORESTS

The initial research undertaken in natural tropical moist forest was scattered. The means were never sufficient to achieve the goals, and research and application were often confused. Experimental plots were established independently in most tropical forest regions, without common guidelines or a similar model, so it was not possible to interpret or compare the collected data. In addition, imaginative efforts often suffered from successive changes in financing and forest policy. Thus, when concern about management of tropical forests increased in the mid-1970s, much of the information on natural forest silviculture was in the form of incomplete and uninterpreted data.

In 1976, the CTFT began establishing a network of experimental plots, based on a series of simple principles:

- only use large plots (several hectares) with as many comparable ones as possible in other places;
- measure simple parameters (e.g. circumference, tree location);
- use statistical means to interpret data.

At the same time, the main objectives focused on the study of:

- stand dynamics and tree growth in relation to simple silvicultural activities (e.g. logging), with attention paid mainly to primary stands;
- regeneration and the influence of silvicultural treatments;
- silvicultural treatments to increase timber yields;
- the development and transfer of technical concepts from experiments to large-scale trials for demonstration and forest management projects.

The experiment network now spans three continents and features national and international exchanges which supply a considerable and steady flow of information. The following sections concentrate on conclusions that can be drawn from growth, productivity, vulnerability, mortality and regeneration data.

Impact of Logging

In the Central African Republic, in a forest with a total standing volume of just over 300 m^3 per hectare (counting trees of more than 10 cm in diameter), harvesting removed three to four trees of over 80 cm dbh, representing 50 to 65 m^3 per hectare. The inclusion of logging damage increased removals to 68 to 95 m^3 per hectare, or 20 to 30 percent of the initial volume. In Guyana and Brazil, where the standing volume may be 310 to 370 m^3 per hectare, the logging of about ten trees of 50 cm in diameter means a harvest of about 50 m^3 per hectare and a total removal of 75 m′ per hectare or 20 to 25 percent of the initial volume. Thus, the direct impact of logging is considerable. Disturbance should be controlled so as not to exceed the threshold of irreversible deterioration. This is especially important, as mortality may be abnormally high during the two to three years after logging. Opening of the canopy during logging may stimulate diameter increment but this affects relatively few trees because of the heterogeneous nature of logging activities.

Effect of Thinning

Thinning was accomplished by devitalising (with or without arboricides) noncommercial or ''secondary" species, which were left standing to reduce cost and damage. Only trees in the higher stratum were devitalised, since these exert the most competition for future crop trees. Logging removes valuable tall trees and thinning gets rid of tall trees that have no commercial value, so the result is to create "younger" stands and stimulate their growth while preserving biodiversity.

The opening of stands by thinning favours the growth of most tree species, especially for small and medium-sized trees. The first results obtained in Côte d'Ivoire were very promising and were confirmed not only by other experiments in a similar environment in Africa but also by those in America in a very different type of forest. In Guyana, for instance, the growth rate of commercial species is increased by 75 percent for stems of 10 to 25 cm in diameter and by 50 percent for 25 to 40 cm trees during the four years after logging. But taller trees subject to less competition may show no response to treatment.

Natural Mortality

Natural mortality is an important mechanism of forest regulation and regeneration. It is an extensive phenomenon but is difficult to quantify and study. It concerns only a few trees per hectare each year but the overall loss in volume may be considerable when large stems are concerned. No connection could be established between natural mortality and the intensity of stand opening: trees die whatever their size and whatever their growth, in both treated and control plots. It is a discontinuous phenomenon with exceptional occurrences and it requires very long periods of observation.

Natural Regeneration

Several experiments addressed stems of 2 to 10 cm in diameter. This research is far from conclusive but suggests that treatment stimulates the regeneration of most species without causing major floristic modifications outside the areas affected by logging. Experiments were unable to establish a link between regeneration and the presence of the species in the over-storey. Estimates in this field will remain provisional for a long time, since a long period is necessary to encompass the successive effects of a single action.

Fire

A fire occurred in 1983 in one of the three projects in Côte d'Ivoire. This accident was used to quantify the scope and the consequences of the damage. On the whole, the areas most damaged by fire were those which had been thinned and exploited and which, because of the abundance of dry standing and fallen trees, were more flammable than the untouched plots. This discovery is not very encouraging for forests that have been modified by humans, but it emphasizes the importance of forest protection as an "everyday matter". Good silviculture is not sufficient; attention must be paid daily to the preservation of investments and capital.

Production

Annual volume production of all commercial species depends on the treatment used. Figures obtained in Africa, for example, show that production doubles during the four years after treatment. As the initial volume of standing trees was 100 to 150 m^3 per hectare, this represents an annual productivity of about 2 to 3.5 percent. Studies in the Central African Republic show similar results despite differences in stand structure, species and treatment.

The volume growth of trees above the cutting limit is more variable. In Côte d'Ivoire, for instance, the gain in production obtained by thinning or logging turns out to be relatively low, between 0.5 and 1.5 m^3 per hectare per year. This may be explained by results stated previously:

— the growth of dominant trees is almost independent of treatment;

— mortality is very variable and may be unrelated to silvicultural activities;

— the gain in volume by recruiting to a commercial size is clearly influenced by the treatment and its intensity. Average-sized trees benefit most from the elimination of competitive trees.

The loss of commercial volume through natural mortality is about the same magnitude as the volume growth of commercial trees. It is the new trees reaching commercial size (60 cm dbh) that are affected by treatment and provide the gain in production quantified here. Volume growth is practically non-existent within untouched forests, so thinning and the mobilisation of resources are extremely important in productive forest management.

CONDITIONS FOR SUCCESSFUL SILVICULTURE

Is the conservation of tropical forests a myth? Some 15.4 million ha are deforested every year! The "developmentalist" and "conservationist" approaches confront each other:

each has limits and constraints. After harvesting, the silviculturist must use the various techniques made available through research to reconstitute the forest resource. The choice of technique will depend, on the one hand, on the management objectives and, on the other, on the constraints inherent in the stands: the remaining trees, the capacity for silvicultural treatment, susceptibility to fire, competing demands for land, etc.

If natural forests are to be conserved, three (non-silvicultural) conditions must be fulfilled as a point of departure before the technical activities of forest management are undertaken:

— at the governmental level, land management plans (supported by legislation that is applied and respected) must prescribe management of stable, demarcated forested areas;

— at the local level, the land allotted to forest should integrate the needs of local people and offer assistance for the development of farmland in the areas surrounding the forest;

— at the resource level, the logger and the silviculturist must work as partners with the understanding that harvesting is part of silvicultural treatment but, at the same time, ensuring that immediate operations and long-term goals are commercially realistic.

Silvicultural Procedure

Assuming that the three above-mentioned conditions are achieved, the first necessary condition from a silvicultural point of view is that harvesting should not exceed production. Inventory and growth modelling techniques provide reliable estimates of productivity and enable informed management. Ideally, individual trees could be identified, quantified and auctioned as standing trees. This approach is used in temperate areas but is far from operational in the tropics, even though it corresponds to the basic concept of the intrinsic value of the standing tree.

The second necessary condition is to plan how trees can be logged. It is obvious that forest harvesting affects the environment, but it also constitutes a silvicultural operation. Properly planned and controlled logging systems are preferable in every aspect: economic, ecological and silvicultural.

The third necessary condition is to enforce the appropriate post-logging measures to stimulate the growth of commercial species by selective thinning of dominant secondary species while being careful to maintain natural biodiversity and sufficient regeneration.

Table 3. Annual volume increase (over 10 cm diameter) of commercial species four years after treatment in two parts of Africa

Treatment	*Côte d'Ivoire*		*Central African Republic*	
	m³ per hectare per year	*Percentage of initial volume*	*m³ per hectare per year*	*Percentage of initial volume*
Untouched	0.7-1.8	0.5-2.0	1.35	0.9
Logged	2.5	25	1.50	2 2.1
Logged and thinned	2.2-3.6	0.2-3.5	2.45	3.2

These three technical conditions are justified in a management plan where the standing tree capital is sufficient to maintain commercial potential but, if the forest is too impoverished for natural regeneration, then plantation establishment becomes justified. The choice of plantation type must take into account the ecological vulnerability and, depending on production goals. the intensity of establishment methods may vary: low for extensive manual methods (enrichment); average for intensive manual methods (taungya, agroforestry systems); and high for mechanised methods.

SILVICULTURE IN BOREAL FORESTS

The northernmost forest cover on earth is called the boreal forest (after *Boreas*, the personification of the north wind in Greek mythology). Encompassing approximately 920 million ha (of which some 820 million ha are potentially exploitable), it encircles the globe and forms a more or less continuous belt through Eurasia and North America. The boreal forest falls mainly within the borders of the Nordic countries (Norway, Sweden and Finland) which have 5 percent, the Russian Federation, which has 73 percent and Canada and the United States (Alaska) which have 22 percent. Boreal forests include vast, remote and sparsely populated wilderness areas, until now beyond the reach of large-scale wood harvesting, as well as areas that have long been an important resource for humans as providers of wood raw material, originally destined for local use but increasingly used for industrial processing at both national and international levels. Currently. boreal forests yield about 500 million m³ of industrial timber per year, equal to 37 percent of the world's total demand, and 45 percent of all coniferous timber harvested.

Industrial Uses

The history of industrial utilisation of boreal forests varies in length from two centuries to nil, depending on factors such as distance to consumers and accessibility. Therefore, some boreal forests harvested today are the result of humans, silvicultural practices in the past while, in other areas, the trees currently being cut were sown centuries ago by nature alone.

In Europe, large-scale utilisation of boreal forests started around the mid-1800s. Initially, the natural forests were exploited in a similar way to mining. This exploitation was the basis for the establishment of mills and surrounding communities. Silvicultural investments aimed at rejuvenating the resource were either low or nonexistent at this time. However, after a few decades or so of forest exploitation, an area's forest resource created by nature alone began to dwindle. Hauling distances to existing wood processing industries increased, as did the costs, but the mills and surrounding communities could not easily be moved, as they represented huge investments and social benefits. What could be done to sustain and improve wood harvesting in nearby areas? This is the transition stage when the role of silviculture is usually recognised.

The ecology of the boreal forests was poorly understood at that time and the selection cutting systems that had been developed and successfully implemented in the more southerly forests of the continent were introduced and widely used. In theory' they seemed ideal: there were no costs for planting and tending and the forest was to be kept in good and productive shape by a skilful use of the axe alone. In many places, however, these practices developed into biological disasters. Without active site preparation, many logged-over areas became covered with dominating aggressive brush vegetation. Only shade-tolerant climax trees that already existed as undergrowth in the previous stand could survive, for example spruce and fir, and thus the forests took on a completely different species composition. Millions of hectares of open, scrub, slow-growing and commercially worthless forests accumulated where the advocates had promised a garden of Eden. In the absence of strong disturbances, either natural or human-caused, the forests' life processes stagnated.

It was not until the 1940s that the failure of selection cutting as a general practice in northern European boreal forestry was commonly understood and accepted. Since then, a formidable restoration programme has turned these spoiled forest areas into a forceful new generation of young, fast-growing trees.

Of the total annual removals of industrial timber, 80 percent or more come from large-scale final felling in natural forests situated on public land in the Russian Federation and Canada. Here, "large-scale" means removals concentrated on the clear-felling of tracts that often cover 50 to 100 ha or more and yield tens of thousands of cubic metres of wood. In the Nordic countries the large forest industry corporations operate their forests in a similar way, although these are second- or third-generation forests as opposed to primary growth. Taking a global view, forestry in the boreal zone is therefore mainly of the large-scale type. Consequently, silvicultural practices used in industrial forestry are characterised by mechanisation and a concern about profitability, cost efficiency, operations control, etc.

Private, small-scale boreal forestry is practiced mainly in the Nordic countries and in eastern Canada, where hundreds of thousands of small woodlot owners share a major portion of the forest. Despite the different situation, silvicultural systems used in these forests do not drastically differ from those of the larger owners. One reason is that the development of alternative silvicultural methods that would better suit the needs of small-scale boreal forestry was given a low priority in the past. Another is the increasing tendency of small private forest owners to sell their wood on stump and let the buyer—industry—carry out any obligatory silvicultural measures.

Silvicultural Systems

Guided by research and experience gained over the past century, forestry in the Nordic boreal forest zone has developed silvicultural systems that are intended to be rational from an operational point of view and that give reasonably safe and satisfactory biological results. During the past decades, the predominant regeneration method has been planting preceded by mechanical site preparation. In the late 1980s, approximately 750 million trees were planted annually on some 3001300 ha of clear-felled boreal forest.

Experience shows that the rate of regeneration failure is comparatively low in plantation forestry planted seedlings give the new forest stand a quick start; the most suitable tree species can be chosen for each site and, often, the seedlings originate from genetically improved seed, measures needed to establish the new stand are simple to plan and administer; the whole regeneration process is finished in a short time and the results are easy to check while corrective measures can be carried out quickly; and, for standardised methods such as planting, expenditures on personnel and other overhead costs can be kept low.

In plantation forestry, seed is a critical commodity. The collection of seed in the boreal forest is a laborious activity and not always possible because of the winter climate. Large forest areas exist in the boreal zone where the summer climate is frequently so chilly that tree seed only ripens on rare occasions. In addition, some conifers in the boreal forest, for example the spruces, are unreliable seed producers. Others, for example fir and larch, shed their seeds only during a short period after the cones have ripened. Therefore, a large portion of tree seed used in plantation forestry today is produced in seed orchards throughout the boreal forest zone. This seed also is produced from selected trees and hence has genetically improved properties.

A scarcity of labour and cost factors have led to the development of tree nurseries which produce seedlings grown in containers suitable for fast and easy planting, either manual or mechanical. These nurseries can be quite big, producing 50 to 70 million seedlings per year. Because of fairly high investment costs, they seldom have an annual capacity of less than 15 million seedlings. Under Swedish conditions, it has been possible to quadruple manual planting performance by using container seedlings instead of bare root seedlings. The planting rate per hour doubled, as did the length of the planting season. Container seedlings are also more suitable for being fed into planting machines than the bare root seedlings.

The importance of appropriate containers is worth mentioning. If grown in inappropriate containers, seedlings of numerous species, particularly pine, can suffer from instability after planting out. Such poor containers were in use initially and hence early pine plantations were negatively affected in the Nordic countries. Since then, new containers have been adopted which allow the roots of planted pines a more satisfactory development.

So-called "natural regeneration" is also widely practiced in the Nordic countries. Some boreal forest sites are not suitable for planting. They are to be found on both the wet and dry extremes on the scale of forest sites. Here, nature often provides the seedlings free if the rejuvenation process is skilfully handled by the forester. However, even on sites that are usually planted, natural regeneration under seed trees of pine can be a good alternative—on the condition, however, that the site can be satisfactorily prepared by scarification or controlled burning and that the local climate is adequately warm to allow the tree seed to ripen. In contrast to pine, spruces are usually not sufficiently wind-hardy to function as solitary seed trees and they are unreliable as seed producers.

In earlier days, frequent commercial thinnings were regarded as an important means to improve growth and value and to reduce natural mortality among pine and spruce trees in Nordic boreal forests. However, research and experience have shown that, in managed stands, foresters can decide within wide limits the amount of wood they want to harvest in the form of thinnings or final harvest. Without influencing the total yield of wood too much, they can choose either: a programme with a few or even no thinnings, harvesting most of the total yield in one single operation at the end of the rotation; or programmes with frequent light thinnings or just a few heavy thinnings. The choice of stand treatment programmes therefore depends more on economic factors, such as the price of wood assortments produced and harvesting costs. It should be noted that the same liberties might not exist in boreal forest zones where other conifers, for example *Larix* spp., are dominant.

Fertilisation with nitrogen has been used during the past decades to maintain the harvest level of wood in some boreal forest areas, for example Sweden and parts of Canada, without risking the long-term supply. In Sweden, tens of millions of cubic metres of wood were produced with the help of aerial applications of urea and ammonium nitrate during a critical period in the 1970s and early 1980s when the resource of forests ready for harvest was temporarily in short supply. By adding extra growth to the growing stock of middle-aged and older forest stands on a large scale, forest owners were able to compensate for the temporary shortage in the age class distribution, which would otherwise have implied a considerable reduction in annual harvests over time. Today, fertilisation is frequently used by the bigger forest owners to improve tree growth after thinning.

Mechanisation has become an important component of boreal silviculture. Mechanical site preparation replaced the labour-intensive controlled burning at an early stage. More recently, there have been attempts at fully mechanised planting and precommercial thinning. Machine planting is a difficult task in some boreal forest areas because of the very rocky morainic soils. Today, however, planting machines with impressive performance capacities are working under these conditions, preparing the planting spot and planting container seedlings in one operation. One such machine in Sweden, working three shifts per day, can plant three million seedlings during the snow-free season.

An important part of boreal silviculture involves matching provenances to site conditions. Many boreal forest tree species can be grown far away from their place of origin in other parts of the boreal forest zone, provided the right provenances are used. In Sweden, for example, American lodgepole pine (*Pinus contorta*, var. *latifolia*) has been planted on more than half a million hectares and, on suitable sites, surpasses the increment of the Swedish pine and spruce by 40 to 60 percent. However, in the development of Nordic silviculture, rules for provenance choice had to be sharpened several times before plantations established far north and/or on high elevations showed acceptable survival. Large areas had to be replanted because of fungal disease on weakened trees, basically caused by climatic maladjustment.

With time it has become obvious that boreal forestry also needs alternative silvicultural systems that are useful for "difficult" forest sites, for the production of special wood assortments and for sites which need special consideration on the grounds of nature values. In the Nordic countries, silvicultural research is now engaged in developing such systems.

Planted and/or thinned trees, given a quick start in life, respond by forming relatively wide annual rings and thick branches. This is in contrast to many trees in the natural forest which were suppressed when young. Such initially slow-growing trees, sought after by the lumber industry for their high-quality sawnwood, will become rare as a result of present day silvicultural practices. Boreal silviculture shares this dilemma of volume yield versus wood quality with many other wood-producing areas of the world. In the Nordic countries, research in boreal silviculture is focusing on the prospects of modifications in order to improve the quality of tomorrow's planted forest.

Silviculture—Investment and Maintenance

In areas where harvest levels depend on yield-sustaining measures, for example in the Nordic countries, silviculture has become a necessary and even unavoidable part of sustained yield forest management. New trees have to be grown and stands tended at least at the same rate as harvesting proceeds. Under these conditions, silviculture tends to become a significant activity and cost item. Its cost must be judged against the gains in economic terms; hence boreal silviculture must be rational and production-oriented. On the other hand, it must also respect the needs of nature conservation and that of other forest users.

In this situation, silvicultural costs can no longer be regarded—as the German forest economist Pressler taught—as isolated investments directed ascertain stands in the forest. Instead, the composition and cost of a forest owner's silvicultural programme must be assessed against the forest's sustained yield level under this treatment only. The programme must be balanced so that the last dollar, rouble or crown spent gives a return that corresponds to the marginal income from wood or other forest benefits produced as a result of this expense.

This somewhat new way of appreciating costs for silviculture more as a maintenance cost than an investment, and hence of taking immediate advantage of the measure's effect on the condition of the whole forest apparatus, can most easily be understood and utilised by private owners or forest-owning enterprises where forest and forest industries are managed in close combination. In Sweden, where big enterprises own 50 percent of the forest, decisions to start intensive silviculture were rapidly taken in the boardrooms of forest-owning corporations when the need was clearly demonstrated half a century ago. Through legislative measures and public financial support, these programmes also spread rapidly to the smaller private forestry operations.

In other countries, for example Canada, where state ownership dominates, forest industries usually extract their wood from leased forest land. The silvicultural programmes needed to reforest the harvested areas are generally defrayed by the forest owner as part of the general budget. As a result, in times of poor economy , silvicultural measures have often been neglected or underfinanced. In Canada the discussions went on for decades before the need for more targeted large-scale silviculture was generally understood and accepted in political circles and the necessary programmes financed and implemented. In the meantime, the forest's general condition suffered. Enormous areas of backlog and huge needs for forest restoration accumulated. Even though improvements have been made, this questionable system of divided responsibility for financing forest operations persists.

Results of the Transformation

Since the start of the new era of boreal forest management in the Nordic countries 40 to 50 years ago, about 40 percent of the total area has been covered with new forests as a result of active silvicultural measures. The middle-aged forest has been thinned and large areas fertilised. As a means to improve forest growth, many owners have saved the better portion of the old forest from final felling and instead directed this activity towards poorer stands. Progress has been monitored through national forest inventories and, overall, the growing stock in the Nordic countries has increased by 23 percent during the past 40 years while the annual increment has been increased by as much as 36 percent, reflecting the creation of a more vital forest capital.

Today we know that the yield capacity of well-tended forest is generally 25 to 30 percent higher than yesterday's predictions. As a result of this as well as other positive changes in forest conditions, wood harvesting budgets have been successively upgraded and are reaching increasingly higher allowable cut levels. Long-term predictions for Sweden's boreal forests show the prospect of increasing the sustainable harvest rates by another 55 percent over today's level. Highly profitable measures, including forest fertilisation and the planting of fast-growing trees such as *Pinus contorta,* have motivated an immediate increase in the annual harvest by 20 percent in large forest estates belonging to some Swedish corporations.

Nature Conservation and Biodiversity

During the past two decades, increasing research resources have been spent on studies of the natural processes in boreal forest ecosystems. As a result, the influence of modern forestry practices on nature values and biodiversity in these forests can be better understood and predicted than before, and they have come under debate in wide circles. As elsewhere, environmentally engaged interest groups are advocating a ban on the utilisation of forests in some areas. The ongoing process of converting natural boreal forests into production forests is criticised. Supposedly more indulgent and environmentally sound forestry practices than those used at present are suggested.

The silvicultural methods used today are, of course, undergoing continued development and modification as a result of changes in society, research findings and practical experience. However, there are some important basics in boreal forestry that should be made clear to "environmentalists".

The challenge that boreal forestry faces today and tomorrow is to reconcile rational and economically successful silviculture with good conservation practices. The economic frames within which boreal forestry, and consequently the societies dependent on them, can exist do not allow the same direction of development as that which characterises the heavily subsidised forestry carried out in many central European communal and state forests, for example. There, the recreation demands of the surrounding dense human population as well as far-reaching nature conservation intentions (partly spurred by pollution damages) have raised the cost of silviculture far beyond the levels of profitability. Boreal forestry must include the cost of nature conservation in its own budget and still be able to survive.

The importance of the boreal forest for supplying humankind with wood-based commodities was stressed earlier. On a global scale, it would be physically impossible to substitute the industrial wood that is currently harvested in the boreal forest by increasing removals elsewhere, and any attempt in that direction would probably bring about other unacceptable environmental effects.

The main portion of industrial wood obtained from boreal forests is still extracted from natural stands, and this situation will persist for a long time to come. For many economic and biological reasons, it is hard to see any advantage in giving up today's harvesting methods and appropriate silvicultural measures in favour of something totally different in these forests. Selection cutting, which has been suggested as a general, "nature friendly" practice in natural boreal forests, would not only be killing for money, it would also repeat the mistakes of the past and seriously damage many boreal forest ecosystems.

Harvesting and silviculture in the boreal forest seldom affect more than a portion of the total landscape. This is an important observation when discussing the impact of large-scale commercial forestry on nature values. The natural mosaic of the terrain can be utilised as part of forest operations aimed at forming a web of small-scale reserves and corridors which are important for sustaining biodiversity and natural ecological processes alongside production-oriented silviculture.

Even where clear-felling takes place, environmental considerations to improve biodiversity must be an important part of planning and execution. The key person to decide which trees to cut and which to save is the machine operator or the person with the power-saw. Operators must be trained to understand the need and be given the responsibility for leaving groups of dead trees and large hardwoods, standing and fallen snags, to save brooks and marshes, etc. Appropriate ecological considerations should also be pan of all ensuing silvicultural work, from reforestation to thinning. In Sweden, practically all forestry personnel and a large number of private forest owners have been trained in nature conservation. Forestry workers and forest ecologists have collaborated to produce illustrated manuals demonstrating the practices to be applied in various forestry activities in order to conserve biodiversity.

A rather unexpected effect of the great forest restoration programmes on both sides of the Atlantic has been the strong positive impact on certain wildlife populations as a result of the changing forest conditions. Clear-felled and planted areas covered with grass, herbs and bushes offer food in abundance for many animals. The formerly fairly small moose populations in the Nordic countries have increased enormously as a result of new silvicultural practices, in fact to the point that winter browsing on young pine trees has become a great problem for forestry. The number of moose has had to be drastically reduced in Sweden: the peak year was 1982 when 175 000 animals were killed. The population, now under control, is still far larger than it has ever been in modern times and allows an annual harvest of more than 100 000 animals.

SILVICULTURE OF MANGROVES

Mangroves are one of the most familiar forms of vegetation occurring in the intertidal zones along sheltered coasts and river banks in coastal areas in tropical and subtropical countries of the world. Mangrove forests are highly productive ecosystems and are a natural, renewable resource. They provide essential goods and services and play a very important role in the lives of coastal communities. Through adaptations such as viviparous germination, the separation of freshwater from salt water and the conservation of freshwater, the ability to strike roots soon after coming into contact with soil and the ability to exchange gases through specialised root systems, mangrove species have been able to deal with a very adverse environment where few other plants would have survived.

However, in many areas of the world, mangrove habitats are being destroyed as rivers are dammed, their waters diverted and the intertidal zone extensively developed for agriculture or aquaculture and generally dried up. Large tracts are being converted to rice fields, industrial and land development and other non-wood uses. In response to the lucrative shrimp export trade, a new breed of small- and large-scale farmers are carving out vast chunks of tidal flats for shrimp farming and pisciculture. Remaining mangrove resources are overexploited for fuelwood and charcoal-making. The depletion of mangroves is also a cause of serious environmental and economic concern for many developing countries, given the pivotal role of this vegetation in coastal protection.

Mangrove Habitat and Characteristics

The mangrove vegetation of the world can be divided into two broad groups. The old world mangroves occurring in the Indo-Pacific region extending from the east coast of Africa to Samoa in the South Pacific. The second group, the new world mangroves, occurs along the west coast of Africa between Mauritania and Angola, in the Americas on the east coast between Barbados and Brazil and on the west coast between Mexico and northern Peru. Indonesia has the largest total area of mangrove forest while the Sundarbans swamp region in Bangladesh and India is the largest single chunk of mangrove forest in the world.

Mangrove vegetation primarily comprises trees and shrubs, with a limited number of palms and lianas. The World Conservation Union's report on the global status of mangroves lists 61 species. Major mangrove species belong to less than 15 families but the most frequently occurring mangroves belong to the Rhizophoraceae, Sonneratiaceae and Avicenniaceae. The Sundarbans, where members of the families Sterculiaceae and Euphorbiaceae predominate, is an exception.

Mangroves grow in conditions where few other plant species can survive. The occurrence of mangroves is contingent on a number of factors and important physiological adaptations which enable the flora to thrive in the exacting environment and which also have silvicultural implications.

The mangrove environment is primarily saline, and the vegetation grows and flourishes by using three different mechanisms which cope with excess salt. The roots of salt-excluding species of *Ceriops, Excoecaria* and *Rhizophora* can absorb only freshwater from the saline water through a process of ultrafiltration. Species of *Avicennia* and *Sonneratia* can regulate the salt content of their tissues by glands in their leaves. *Xylocarpus* sp., *Lumnitzera* sp. and *Sonneratia* sp. deposit salt in older leaves, roots and bark. Mangrove plants also display features similar to those in desert plants which tend to conserve water.

Mangroves occur in areas where strong wave actions are absent. The most extensive growth of mangroves can be seen in estuaries of rivers and protected lagoons and coastal lakes. Mangroves occur in areas of high humidity and their luxuriant growth is often associated with a high rainfall. Minimum air temperature and seasonal variations in temperature are important in the growth of mangroves. As cited by Chapman, the best mangrove growth and development occurs where the seasonal temperature variation does not exceed 10°C and where the air temperature in the coldest month is higher then 10°C.

Mangrove soils are characterised by high salt and water, low oxygen and high hydrogen sulphide contents and contain a high proportion of humus. The best growth and development of mangroves takes place on alluvial and muddy soils which are generally formed by the deposition of water-borne soil particles. Mangrove soils are mostly anoxic except for the surface layer in which roots spread. As a result, mangroves generally have shallow root systems and therefore cannot withstand strong wind and grow better in a sheltered habitat.

Mangrove seeds and propagules are dispersed exclusively by water and their distribution is therefore greatly influenced by tides which carry them both upstream and downstream. The tidal range, together with the topography, of an area regulates the lateral extent of mangrove development. The greater the tidal range, the greater the vertical range available for mangrove communities. Tides also bring about changes in the salinity concentration of water in mangrove areas. Mangrove forests have an efficient mechanism for natural regeneration, particularly in areas within mangrove stands where site degradation has not taken place. Because of the daily or periodic inundation by tidal water as well as the deposition of water-borne soil particles, the forest bed is soft and always suitable for regeneration and normally does not require any site preparation.

One other notable phenomenon in a large number of important mangrove species (*Rhizophora, Ceriops, Bruguiera, Kandelia and Nypa*) is that their seeds develop into seedlings while they are still attached to the mother tree. This phenomenon is known as vivipary where the embryo ruptures the pericarp and grows beyond it. In a second group, which includes species of *Aegiceras, Laguncularia, Pelliciera* and *Avicennia,* the embryo, while developing within the fruit, does not enlarge sufficiently to rupture the pericarp. This has been termed as cryptovivipary by Hutchings and Saenger. Species of *Excoecaria, Sonneratia, Heritiera* and *Xylocarpus* are not viviparous. However, all of the above-mentioned species produce buoyant seeds which are dispersed exclusively by water and are capable of reaching any corner of the forest inundated by tidal water.

Most seeds that fall off mother trees during low tide stick to soft mud and quickly strike roots. In the case of viviparous seeds, the adventitious roots already present in the hypocotyl emerge and anchor the seedlings. However, in all cases the deposition of silt by subsequent high tides helps the seeds or seedlings to secure a better hold.

In instances where seeds or seedlings drop during high tides, they continue to float in the water until they come into contact with a soil substrate and strike roots. It is interesting to note that most mangrove seeds retain viability for a long period while still in the saline environment, but lose viability soon after they are removed from that environment. Profuse natural regeneration seems to occur in areas under mangrove formations. In many instances, advance growth establishes itself and waits for an opening in the canopy to emerge. On the other hand, radical changes in site conditions through the clearing of existing vegetation or alterations in the water regime seem to affect the regeneration process adversely. Dry soil in open areas often does not favour natural mangrove regeneration and is invaded by the undesirable species of *Achrostichum*.

Selection System

The selection system is practiced in Pakistan, India, Bangladesh and Myanmar where trees above certain predetermined diameters are harvested from the annually stipulated coupes. The Sundarbans was managed under a selection system in 1892-1893 when the first management plan for the forest was implemented, and the modified selection system, which was fine-tuned and adapted during the first three decades of this century, has proved to be very appropriate for the sustainable management of the forest. The depletion that has occurred in the growing stock has resulted from faulty management decisions rather than silvicultural practices.

The mangroves of the Sundarbans are managed under a selection-cum-improvement silvicultural system. Separate annual coupes are laid out for timber, fuelwood and pulpwood extraction. All types of harvest are carried out in a 20-year cycle where a single harvest operation is carried out once in each 20-year period.

The specific process for *Heritiera fomes*, which is the principal timber species in the forest, involves the laying out of the annual coupe, the marking of sound trees above a predetermined exploitable diameter provided their removal will not create any permanent gap in the canopy—the harvesting of timber followed by the removal of dry lops and tops and, finally, improvement felling of all deformed trees and the thinning of dense stands. This conservative opening of the canopy ensures an adequate regeneration of this species, which thrives in partial shade in its early stages, while discouraging the regeneration of strong, light-demanding and economically less desirable species.

Excoecaria agallocha is exclusively used as pulpwood and matchwood. All trees above the exploitable diameter are harvested in a single operation. In the case of *Sonneratia apetala*, all trees above 30 cm in diameter are removed, provided such removal does not create any gap in the forest canopy. Within annual coupes where *H. fomes, E. agallocha, Ceriops decandra* and other mangrove species have been established as understorey, clear-felling of all *Sonneratia apetala* trees is carried out to assist in the establishment of a more valuable crop. *S. apetala* seedlings do not establish under a mature crop of the same species. *C. decandra* is used both for poles and fuelwood. Poles in a coupe are removed following selection felling rules as described in the case of *H. fomes*. During the fuelwood harvest, at least one healthy shoot is left in each branch.

REFERENCES

Adegbehin, J.O. & Nwaigbo, L.C. 1990. Mangrove resources in Nigeria: use and management perspective. *Nature Resour.*, 26(2): 13-21.

Aksornkoae, S. 1993. *Ecology and management of mangroves*. Bangkok, IUCN.

Chapman, VJ. 1975. Mangrove biogeography. In G.E. Welsh, S.C. Snedaker & HJ. Teas, eds. *Proc. International Symposium on Biology and Management of Mangroves*, Vol. 1. Gainesville, USA, University of Florida Press.

Chapman, V.J. 1976. *Mangrove vegetation*, Vaduz, Liechtenstein J. Cramer.

Chapman, V.J. 1977. Introduction. In *Ecosystem of the world*, Vol. 1. *Wet coastal ecosystems*. Amsterdam, Elsevier.

FAO. 1989. *Management of tropical moist forests in Africa*. FAO Forestry Paper No. 88. Rome.

FAO. 1993. Timber plantations in the humid tropics of Africa. FAO Forestry Paper No. 98. Rome.

FAO. 1994 *Management and conservation of closed forests in tropical America*. FAO Forestry Paper No. 101. Rome

5

Agroforestry for Environmental Management

Multipurpose tree species grown in combination with agricultural crops and known to have direct bearing on rainfall erosivity, soil erodibility, and soil surface protection. Just to enter plant trees in croplands and especially on clean cultivated steep slopes may sometimes aggravate soil erosion. Although the volume of rain reaching the ground is reduced somewhta by interception of tree canopy, there is an increasing problem from leaf drip which develops a drop size larger than uninterested rain drops and creates a fall distance greater than that intercepted by ordinary crop covers. Both drop size and fall distance will add to the kinetic energy hence increase soil detachment and erosion. A dense canopy of low tree of shrubs, such as provided by coffee or tea bushes, reduces erosivity, although the shade trees in plantation increases it. In spatial mixed agroforestry system, therefore any such effect will depend on the height of the canopy.

In spatial zoned systems, including hedgerow intercropping, the canopy is usually low but it is not vertically above the cropped land. Thus, maintenance of a ground surface cover of 60% or more, formed by any combination of living herbaceous plants with plant litter, has a high potential to reduce erosion, and should be the primary objective in agroforestry design. It is stated that the effectiveness of agroforestry in controlling erosion should rest on a foundation of experimental measurement of erosion rated under actual agroforestry systems. Management practices are the main deciding factors and determine the merits and demerit of various agroforestry practices whether they are controlling or aggravating the soil erosion. Hedgerows of Leucaena and Gliricidia at two and fourm spacings were compared with no tillage and conventional ploughing without hedgerows. The results showed that the mean rates of soil loss over two year (t/ha/year) were 8.75 under ploughing, 6.95 under hedgerow inter cropping (mean of two hedge species, two spacings) and 0.02 under no tillage. Reduction of run off and nutrient losses followed the same pattern thus, although hedgerow intercropping was not effective as no tillage, it reduced soil and nutrient losses and run off, to well below acceptable limits.

On steep slopes in Colombia (humid climate, rainfall 4000 mm) one year record soil losses of 23-38 t/ha/year under maize reduced to 13 t/ha/year (on both 49 and 75% slopes) by hedgerows of Gliricidia.

NFT IN AGROFORESTRY AND IMPROVING SOIL FERTILITY

Nitrogen fixing trees and shrubs, growing within practical agroforestry systems, are capable of fixing about 50-100 kg N/ha/year. The nitrogen returned in litter and pruning may be 100-300 kg. N/ha/year, fourthly derived by recycling of rertiliser nitrogen. The major role of trees it to improve the efficiency of nutrient cycling, the mechanisms are uptake from Lower horizons reduction of leaching loss by tree root system, balanced nutrient supply and improvement in the ration between available and fixed minerals. For a tree leaf biomass production of 4000 kg. DM/ha/year. The potential nutrient return in litter, as kg/ha/year, is the order of 80-120 kg for Nitrogen 8-12 kg for Phosphorus 40-120 kg for Potassium and 20-60 for Calcium. Pruning of green parts of MPTs under agroforestry systems and their incorporation in soils before leaf fall is beneficial over fall as nutrient content in living leaves are higher than in litter. Deciduous trees typically trans locate nutrients from leaves to perennial organs well before leaf fall.

Table 1. Rates of erosion in tropical forest and tree crop system

Land use system	*Erosion (t/ha/year)*		
	Minimum	*Medium*	*Maximum*
Multistory eta garden	0.01	0.06	0.14
Natural rain forest	0.03	0.30	6.16
Shifting cultivation fallow perion	0.05	0.15	7.40
Forest plantation undisturbed	0.02	0.58	6.20
Tree crops with cover crop or mulch	0.01	0.75	5.60
Shifting cultivation cropping period	0.40	2.78	70.05
Taungya cultivation period	0.63	5.23	17.37
Tree crops clean weeded	1.20	47.60	182.90
Forest plantations burned/litter improved	5.92	53.40	104.80

AMELIORATION OF ENVIRONMENT BY MPTS UNDER AGROFORESTRY SYSTEM

The indiscriminate felling of trees to expand foods production and the harvesting of forest products in recent past have begun to impact directly on the economic and environmental health of many countries particularly in the developing one. Pressure on temperate forests would have substantially following centuries of clearing for agriculture. Unfortunately, chemical stresses from air pollution and acid rain today place a substantial share of European forests at risk. Trees covering some 31 million ha in Central and Northern Europe are showing signs of damage linked to air pollution.

Conversion to cropland is by far the leading direct cause of the less of forest in tropical countries. Popullation growth inequitable land distribution and the expansion of exports agriculture have greatly reduced the area of cropland available for subsistence farming, forcing many peasants to clear virgin forest to grow food. These displaced cultivators often follow traditions of continuous cropping that are ill suited to fragile forest soils. Eventually the soils get so depleted that peasant colonists must clear more forest to survive Population pressures have also transformed fuelwood collection into an unsustainable practice, particularly in Africa and Asia.

Fortunately, reforestation/agrooforestry is also proceeding somewhat faster than official estimates indicate Spontaneous tree planting by villages around farm fields, as wind breaks, or along roadways is frequently not counted. Indeed agroforestry statistics typically ignore trees outside of forests, even though in many areas they are primary source of fuelwood, fodder and building materials.

Carbon Cycle and Afforestation

MPTs play acrucial role in the global cycling of carbon. The earths vegetation and soils hold some 2x10 milion tonnes of carbon, roughly triple the amount stored in the stamosphere. When trees are cleared or harvested, the carbon they contain as well as some of the carbon in the soil is oxidised and released to the air, adding to the atmosphere store of CO_2 This release occurs rapidly it the trees are burned but slowly if the decay naturally.

MPTs and Global Warming

Although the cutting and bushing of MPTs clearly adds significantly to the CO_2 buildup, the way ramaining forests worldwide respond both to the buildup itself and to the resulting planetary warming could boost an even greater influence on the earth future climate.

Higher CO_2 levels might enhance the growth of trees, causing them to remove more carbon from the atmosphere, which is turn would dampen the washing Greenhouse operators take advantage of this rertilising effect to boost crop production by setting the CO_2 concentration in greenhouse air two to three times higher than in the natural atmosphere, however, no comparing evidence suggests that natural forest would respond in this way.

If respiration exceeds photosynthesis for an extended period trees would stop growing and ultimately die.

Better use of Sewage through Tree Plantation

In order to combat the disastrous effect of sewage water following points may be taken into account:

1. Sewage should not stagnate so as to produce foul smell and act as breeding ground for mosquitoes.
2. It should not percolate down to contaminate ground water and raise water table thus, water should either evaporate or be consumed.
3. It should not increase the salinity of the soil and/or make it unproductive by means of accumulation of toxic elements.
4. Any vegetation raised at such places should not be directly consumed by human beings.
5. It should be economical rather revenue generating.

In the light of above, technology involving forestry/agroforestry has been developed wherein waterlogging, stagnation and thus pollution is eliminated by utilising the irrigation and nutrient potential of these waste waters, right in the area of their production.

The MPTS plants use the water and nutrients of sewage for their growth and boidrains excess water without any harmful effect on the environment Each tree acts as small biopump absorbing water from the soil and releasing it in the environment through transpiration. Normally, the applied effluent disappears within 12 to 18 hours resulting no water stagnation which in turn lowers production of foul smell eliminating breeding place for mosquitoes and checks rise in water table. There is no adverse effect of sewage disposal on toxicity or efficiency of any nutrient, heavy metal or salinity stresses on plants.

This techniques utilises the entire biosystem as living filter for supplying nutrients to soil and plant irrigation renovates the effluent for atmospheric recharge and ground storage. It also conditions the humidity of the environment and conserves the resources right in th area of their production.

It builds up the soil ferfility with respect to soil available P,N,K 20, C and micronutrients. It decreases the soil pH from highly alkaline to neutral levels without significantly building up the salinity. At CSSRI, Karnal the results showed that the pH of the plots receiving sewage water daily dropped from 8.40 to 7.4 in 4 years. At Nilokheri (near Karnal), where pH of the original soil was 9.60 it dropped to 8.10, simply that it also ameliorates the sodic soil.

Further, as forest plants are to be used for fuelwood, timber or pulp, there is no chance of pathogens, heavy metals or organic compounds entering the human food chain system point which is a limiting factor when vegetables or other crops are grown with sewage.

Suitable Tree Species

Though most of the plants are suitable for utilising the effluent, yet, those tree species which are fast growing can transpire high amount of water and are able to withstand high moisture content in the root environment are most suitable for such purposes. Eucalyptus hybrid which has the capacity to transpire large amount (80 litre/DAU) of water, and remains active throughout the year. Other species suitable for this purpose are Poplar and Subabul.

Out of the tree species, Poplar is the most responsive in utilising sewage however, being deciduous it remains dormant in winter and thus can not bio drain effluent during winter months. Keeping this in view, Eucalyptus seems to be the best choice, for such purpose. However, if area is available and the volume of effluent is small combination of Poplar with Eucalyptus is the best proposition.

Developments in the field of agroforestry has been rapid since early 1980 Today, there is tlittle need to promote agroforestry to a doubtful scientific and development community. The rapidly expanding interest in agroforestry in recent years, witnessed by a myriad of research and development activities, leaves no doubt the agroforestry as an approach to land development is now accepted by most, if not all disciplinary scientist and development speciallists. Increased concern at the highest International Policy levels about the sustainability of agricultural development. In the light of the apparent rapid depletion of the natural resource base, has brought agroforestry even further even further into the limelight.

Trees and shrubs, if properly chosen and managed, have a potential to cope up with disaster and create a congenial environment in addition to conserving the soils productive capacity Soil conservation is not seen in its traditional, narrow sense of preventing water and wind erosion, but in the broader and much more important sense of maintaining soil fertility. We know that it is impossible to reduce the soil loss to zero Limits to be set as targets for the design of land use system. There need to be set low enough such that there will be a serious or progressive decline in crop production yet high enough to be realistically achievable. The concept of tolerable erosion or soil loss tolerance has often misleadingly used. It originated at the time when erosion was viewed primary as physical loss of soil material, which is really disastrous and poses environment degradation and creates impediments in the developmental activities.

Agroforesty is an inter disciplinary, multi-sector approach of land use. Its prime objective is over all optimisation and to protect the environment and maintain the ecological integrity. Agroforestry is primarily the technology of using perennial vegetation in combination with seasonal or perennial field crop, fodder, or other crops of economic value in agriculture. In the developed agroforestry systems, there are three components i.e. the woody perennials, the agricultural crops and the animals. However, in all the agroforestry system the presence of the tree component is must and it plays the dominant roles- the trees have two major roles the productive role (fuel, fodder, food, fruit and fertiliser) and the service roles.

Trees have established themselves as insurance against natural disasters like environmental pollution. floods and droughts. Agroforestry also extends greater diversity for wildlife and landscape improvement. Development activities of agroforestry leads to control of erosion either through barrier approach or by cover approach, maintenance of soil fertility by way of organic matter maintenance, nitrogen fixation nutrient cycling and augmentation of nutrient uptake i.e. trees associated mycorrhizal system However, in defth studies are needed over a range of environmental conditions for improving soil fertility and reducing the dependence on chemical fertilisers. Incorporation of multipurpose trees species with crop can also enhance groung recharge.

Agroforestry to Conserve Atmospheric Carbon Dioxide

The accumulation of greenhouse gases in the atmosphere, particularly carbon dioxide (CO_2), is projected to alter the earth's climate. The potential role of forests in carbon sequestration has recently been evaluated by a number of authors. Although they are preliminary, these analyses suggest that forest conservation, establishment and management as well as agroforestry could contribute to global carbon sequestration and conservation while providing goods and services in local communities of many countries. At the same time, the authors of these analyses agree on one critical point: forest carbon sequestration options alone will not solve the problems related to greenhouse gases. Addressing the climate change issue on a global scale will require complex adaptation and mitigating measures affecting all social and economic sectors. Moreover, it is clear that any forestry-based responses should represent a sound policy that is independent of the predicted global warming, and should produce net benefits in addition to those that may ultimately arise in the climate change context.

In 1989, representatives from 67 countries at the Dutch Ministerial Conference held in Noordwijk, the Netherlands, discussed the role of world forests as carbon sinks. An agreement was reached to recommend to the Intergovernmental Panel on Climate Change that the goal of a net increase in world forest area of 12 million ha/year be adopted by the beginning of next century. This goal would be achieved through the conservation of existing forests, reforestation of degraded forest lands and afforestation of marginal, agricultural, pasture and savannah lands. As part of the Global Change Research Program of the United States Environmental Protection Agency (USEPA), an assessment was initiated in 1990 to evaluate forest establishment and management options to sequester carbon and reduce the accumulation of greenhouse gases in the atmosphere. Three specific objectives were formulated:

- identify site-suitable technologies and practices that could be utilised to manage forests and agroforestry systems to sequester and conserve carbon;
- assess available data on site-level costs of promising forest and agroforestry management practices;
- evaluate estimates of technically suitable land in forested nations and biomes of the world to help meet the Noordwijk forestation targets.

Approaches to the Assessment

The first step in the USEPA assessment was the development of a global database, storing regional and national data from 94 nations in three major categories: forest growth and conservation resulting from forest establishment and management practices; the associated costs for each management practice; and the area of land potentially suitable for each practice. Information for the database was gathered from international scientific and technical literature published over the past ten years and from professional forestry contacts worldwide. Although more than 90 nations are represented in the database, the assessment focuses on 16 key countries (Argentina, Australia, Brazil, Canada, China, Congo, Germany, India, Indonesia, Malaysia, Mexico, New Zealand, South Africa, United States, the former USSR and Zaire).

Carbon Storage Estimates

Data on forestry growth and yields provided the basis for estimates of potential carbon storage. Stem wood volume was converted to whole tree biomass, which was assumed to have a 50 percent carbon content. Whole tree biomass included roots, but not soil carbon, detritus or humus. Annual carbon accumulation and storage were derived by summing the carbon standing crop for each year in the growing cycle or rotation and dividing by the length of the rotation. This estimate of the average amount of carbon on site over a rotation applies to a series of rotations if we assume that the system is sustainable and there is no yield reduction in later rotations.

It is noteworthy that this approach also assumes that at harvest, or shortly after, most stored carbon returns to the atmosphere. This is based on an estimate that at least 40 percent of the total carbon in tree biomass is in leaves and branches, which are either burnt or decompose quickly after harvest. Of the remaining 60 percent of harvested carbon, in most processing operations, less than half of the total volume remains in final products. Therefore, more than 75 percent (and by some estimates as much as 90 percent) of stored carbon returns to the atmosphere quickly after harvest. This in no way negates the opportunities presented for long-term storage of carbon in finished wood-based products; rather, it emphasizes the need for more efficient harvesting and processing operations.

Cost Estimates

Cost estimates were based on implementation costs per hectare for various forestry and agroforestry practices. Implementation cost reporting methods vary but generally include site preparation, stock costs and planting labour plus supervision. Thus, the estimates had three important shortcomings: the cost of land was not included because of obvious calculation difficulties; the analysis did not include annual or maintenance costs; neither social nor political variables affecting costs of management practices were considered. However, an analysis of annual maintenance costs and associated benefits is now being undertaken by USEPA.

Because forests are renewable resources, the costs of establishing them are recurring costs. The assessment used standard economic discounting techniques to compute the present value of a series of successive future costs. Published cost data were converted to US dollars and used to calculate the present value of all establishment costs for a 50-year period. The assumption of 50 years is arbitrary, but one that seems appropriate for a global programme of forest management that would require significant operational start-up time as well as a time-lag before planted trees reach their full potential.

It is particularly important to emphasize that the costs presented are gross, i.e. they do not account for financial or social benefits that result from the initial investment or the production of useful products. To calculate net costs, the present value of future revenues as well as any subsidy inputs would have to be subtracted. As methods are developed and all benefits can be fully accounted for, it is likely that the value of the benefits will compensate for, if not exceed, a large portion of the costs.

Analysis of Promising Management Options

The conservation of world forests, particularly the reduction of deforestation in the tropics, is estimated to be one of the greatest potential contributions to offsetting the buildup of atmospheric CO_2. The causes of deforestation have been extensively reviewed; in order of importance, they broadly include shifting cultivation, the clearing of forests for animal grazing and exploitive tree harvests. The replacement of shifting agricultural practices with agroforestry could potentially reduce deforestation and, consequently, reduce carbon emissions from tropical regions.

Direct evidence of this potential is limited, but one research study indicated that a low input agroforestry system, involving the rotation of acid-tolerant crops, produced agricultural products on a single hectare equivalent to the volume normally produced on 5 to 10 ha under slash-and-burn agriculture. One hectare of closed tropical forest can contain up to 220 tonnes of carbon (t C) most of which, when burnt, is released into the atmosphere. Therefore, for each hectare of agroforestry established on deforested land in the tropics, perhaps as much as 2200 t C could be prevented from going into the atmosphere. Of course, agroforestry occurs in many forms which may be more or less productive.

Regarding the expansion of world forests, an analysis of the database information on potential carbon sequestration values according to different management practices helps to determine which are most promising. The five most promising practices in terms of median value for carbon sequestration are discussed below in descending order.

Natural regeneration in the tropics. This practice has a median value of 195 t C/ha over a period of 50 years. This is the highest median value found in the database and it possibly reflects the elevated biomass productivity rates of natural ecosystems in the humid tropics. However, this practice currently has only three references in the database and needs more verification.

Afforestation in temperate latitudes. The median value is 120 t C/ha, a high value which probably reflects the high growth rates of plantations established on lands removed from agricultural production. Although these lands may have been only marginally productive in terms of agricultural crops, for forest plantations they have often represented particularly high-potential sites.

Agroforestry in the tropics. The assessment shows a median value for this practice of 95 t C/ha. Tropical agroforestry has been under intensive study for only a decade and, in comparison with reforestation, there are consequently fewer data while those that are available may be skewed towards high-end results. Nonetheless, the values for agroforestry are encouraging because this practice will certainly be important from the standpoint of supporting local populations.

Reforestation in the tropics. This practice has a median carbon sequestration value of 65 t C/ha. The 136 entries in the database lend considerable weight to the validity of these sequestration values while the high ranking of this practice supports the conclusion, often advanced, that reforestation in the tropical latitudes has great potential.

Reforestation in the temperate latitudes. At a median value of 56 t C/ha, this approach is the fifth highest on the list of promising practices for carbon sequestration. The estimates are based on 212 entries, the largest number in the database. Reforestation, as compared to "afforestation", described above, usually involves less dense stands on poorer soils, thus accounting for the difference in sequestration rates.

It is noteworthy that silvicultural practices had the lowest median value among management practices in all three latitude regions. Silvicultural treatments, such as thinning and fertilisation in plantations, will almost certainly play an important role in adapting forests to the predicted changes in climate. However, by themselves, such treatments will not produce an increase in forest area and their potential to offset the buildup of atmospheric CO_2 appears to be low.

Cost of Forest Management Options at Site Level

The site-level cost of implementing promising carbon sequestration options was one of the primary objectives of the overall assessment. Implementation or initial costs of forest establishment and management generally appear to be lowest in boreal regions. As management intensity increases in temperate and tropical regions, initial costs per hectare escalate accordingly.

For the boreal forest system, natural regeneration and artificial reforestation could sequester carbon most efficiently at a cost of US$93 to $324/ha. At sequestration values of about 17 t C/ha and 39 t C/ha for a 50-year period, the initial costs for the two practices are $5 ($4 to $11) and $8 ($3 to $27)/t C, respectively.

In temperate regions, reforestation, afforestation, natural regeneration and silvicultural practices offer the least expensive opportunities for sequestering carbon. Artificial reforestation can cost $357/ha at a sequestration rate of 56 t C/ha. Carbon is stored at an initial cost of $6 ($3 to $29)/t C depending on site conditions, tree species and management intensity. Afforestation can store about 120 t C/ha at a cost of $259/ha or $2 ($0.22 to $5)/t C. Natural regeneration can be very inexpensive at less than $$10/ha or at 9 t C/ha, the cost is less than $1 ($0.01 to $0.43)/t C.

The widest range of costs was reported for forest carbon conservation or sequestration options in tropical latitudes. Natural regeneration, short-rotation fuelwood plantations and agroforestry systems can all be established for less than $1000/ha (50-year cost basis). Reforestation and agroforestry can sequester carbon at less than $10 ($2 to $26)1t C because of the high sequestration values. Intermediate silvicultural treatments stimulate productivity and can sequester carbon at approximately $500/ha or $8.50 ($1.50 to $36)/t C at a sequestration value of 59 t C/ha.

Overall potential and cost

A marginal cost analysis integrated data on carbon storage, establishment costs and land area. Forest management practices and their associated potential land areas were ranked in ascending order from the least to the most expensive per tonne of stored carbon. Assuming that it is most rational to begin by storing the least expensive carbon, we can accumulate both carbon and costs as we go through the ranked list, adding ever more expensive practices. This approach showed that the marginal cost of storing 45 to 65 gross tonnes (Gt) of C (1 Gt = 10^9 t) would be about \$3/t C with a total cost of \$135000 million to \$195000 million. At more than 70 Gt C the marginal cost escalated sharply to over \$100/ t C. Storing 45 to 65 Gt C would require from 400 to 950 million ha.

For a clearer perspective on these cost estimates, it is noteworthy that a recent study by the United States National Academy of Sciences categorised greenhouse gas mitigation options that cost less than \$33/t C as low-cost. A policy of taxing carbon emissions from fossil fuels could cost as much as \$100/t C. A subsidy of \$0.02/kwh to promote the use of non-fossil fuel energy sources could cost from \$75 to \$150/t C.

Land suitability

The picture of land suitability and availability is clouded by a lack of reliable data and complicated by economic, social and land-use issues. Nonetheless, there appears to be a large area of land in the world that is available for and would benefit from tree-planting. For the tropics alone, Grainger estimated that there are 621 million ha which are technically suitable for the establishment of tree plantations; Houghton, Unruh and Lefebvre estimated that there are 579 million ha available for plantation establishment, 858 million ha for natural regeneration and regrowth and 500 million ha for agroforestry. Trexler attempted to factor in social and competing land-use constraints and estimated that, for tropical Africa and Asia, there are 46 million ha available for plantation establishment, 163 million ha for natural regeneration and 102 million ha for agroforestry.

Approximate areas of suitable land in the boreal regions appear to be 100 million to 200 million ha; and in temperate regions 200 million to 300 million ha.

A conceptual approach to the Noordwijk target

Can forest management and agroforestry practices throughout the world undergo large-scale expansion soon enough to be of significant aid in offsetting the buildup of atmospheric CO_2? Using the forest goals of the Noordwijk Declaration as a framework, the stepwise approach presented below suggests a positive answer to this key question.

The Noordwijk proposal is to achieve a net annual increase of 12 million ha of *new* forest area over world deforestation. Based on current estimates of deforestation of about 17 million ha/year and projections that this rate could reach 30 million ha/ year by 2045, implementation of the Noordwijk. Declaration would require more intensive forest practices on about 30 million to 40 million new ha/year. The following discussion assumes a target of 35 million ha/year for the period 2000 to 2040.

An "easy first" paradigm

With regard to a global net forestation plan, an "easy first" paradigm is suggested. Using this approach, plans make allowance for the programme to start where the obstacles are minimal. Simultaneously, research and negotiations can be undertaken to resolve obstacles. The "easy first" approach is not suggested as a total solution. It is overly simple in relation to the many day-to-day resource constraints as well as the specific social, economic and political issues within individual nations. But the point that the "easy first" approach attempts to illustrate is that all financial commitments, socio-political agreements, technical expertise, etc., although they are critical, do not have to be fully in place before work can start.

In an analysis of forest management practices with potential to slow deforestation and increase forest area, Andrasko suggests three strategies: *i*) maintain forest area; *ii*) reduce loss of forests; *iii*) expand forest area. Following are some possible approaches to the 35 million ha/year target using this scheme.

Agriculture plus forest conservation. As a start, increasing sustainable agricultural practices in the tropics would contribute significantly to strategies *i*) and *ii*). Sanchez estimates that for every hectare placed in sustainable soil management for agriculture, 5 to 10 ha of tropical forest could be conserved. Ross-Sheriff and Cough suggest that it is not unreasonable that a rate of about 1 million ha/year could be achieved in ten years, starting with a first-year level of 50000 ha. If the ratio of sustainably managed hectares to reduced deforestation averages 7.5:1, then 7.5 million ha of the 35 million ha goal could be achieved by the year 2000. Of course, the actual outcome will also depend on the future path of population growth, but fivefold to tenfold increases in agricultural output would have a major impact.

Other approaches cited by Andrasko for strategies *i*) and *ii*) are forest reserves; extractive reserves; natural forest management; and the increased use of pastures (increasing carrying capacity through improved management practices, including seeding and fertilisation, silvipastoral practices, reclamation of degraded pastures, etc.).

Assuming that the potential of these approaches could offset another 2.5 million ha/year of deforestation by the turn of the century, perhaps 10 million ha of the 35 million ha goal could be achieved by these practices.

Agroforestry. For agroforestry, Houghton, Unruh and Lefebvre estimate that about 500 million ha of former forest land might be available (60 million ha of degraded forest land; 38 million ha of woodland; 402 million ha of grassland).

If implementation of agroforestry could increase to an annual rate of 1 percent of the total, then as much as 5 million ha of new tree-growing land could be added each year.

Reforestation and afforestation. To increase forest area, both reforestation and afforestation are required at increasing rates. In this discussion, it is assumed that the annual rates of reforestation and afforestation of marginal lands can be increased by the year 2000 to 2 million ha/year in the boreal latitudes and to 3 million ha/year in both the temperate and tropical latitudes.

Restoration of degraded lands. The tropical latitudes have the primary potential for restoring degraded lands. Grainger estimates that 621 million ha of degraded tropical land have the potential to support forest plantations. Houghton, Unruh and Lefebvre conclude that about 580 million ha of degraded but ecologically suitable lands could he put into plantations. Assuming, therefore, that 600 million ha is a reasonable approximation of the amount of degraded land in the tropics today, the suggested Noordwijk goal for restoration is about 10 million ha/year starting in 2000. In India, approximately 1 million ha of degraded land is reclaimed each year. For the boreal and temperate latitudes, it is also assumed that each of these areas would have approximately 1 million ha of degraded land that could be restored annually.

In total, therefore, these estimates reflect a large enough potential pool of land for expanding the world's present forest area by about 20 million ha annually. Reforestation and afforestation would account for 8 million ha and the restoration of degraded land for 12 million ha.

National forestation goals

Assuming an international commitment to achieving the Noordwijk goal, nations with forest resources would then need to select from a list of forest management options and make their contribution to the overall target according to the size of their respective forest land base while, most importantly, adopting the "easy first" concept within the framework of an international agreement designed to ensure equity for all nations.

Limitations and Needs for Future Research

The current assessment of biological and cost information from more than 90 countries worldwide represents the first significant attempt to develop a bottom-up global analysis. However, before practices can be widely and successfully implemented, consideration must be given to the array of possible economic and socio-political constraints.

A key concern is land availability. Data on the suitability of land for reforestation and other forest management practices are highly variable and this is a serious constraint. It is also important to draw a distinction between land that is technically suitable for reforestation and forest management and land that is actually available. Because of population pressures and demands from competing land uses, much of the land that appears to be technically suitable for forest management may not be available.

Moreover, the degree to which present day forest management practices can simultaneously accomplish objectives as different as the maintenance of global forest area, sustainable economic development and the conservation of biodiversity, among others, is still untested. Research by USEPA and other national and international groups is recommended on the following topics: biomass productivity; costs of the full range of forest management practices; benefits of forest management practices; cost-benefit analysis; risk, uncertainty and constraints in forest and tree establishment and management; land use and tenure; and related issues. Overall, the biological opportunity to conserve and sequester carbon in the terrestrial biosphere, especially in forest systems, appears significant. With careful planning and implementation, management practices useful for this carbon benefit would appear to have potential to provide food, water, wood and other basic human needs. Although implementation costs seem modest, a primary research objective is to place reliable values on all possible forest benefits. Benefit values would allow a clearer determination of the net costs of sequestering carbon through forest management and agroforestry systems. On a global basis, these benefit values, together with more accurate estimates of land availability, will ultimately lead to a more definitive assessment of promising forest and agroforestry practices for sequestering and conserving carbon.

SERVICE FUNCTIONS OF AGROFORESTRY SYSTEMS

The formal study and promotion of agroforestry systems (AFS), a method of land management used since time immemorial throughout the "old" as well as the "new" worlds, started at the end of the 1970s. Initially the focus was on the description, possible biological and socio-economic advantages as well as disadvantages and the inventory of traditional AFS, mostly in the tropics. This was followed by evaluations of productivity of both existing and novel AFS and more recently studies on the interactions between the component species with a view to improving management and profitability. At the end of the 1990s, increased international concern about environmental issues led to new treaties and emphasis on the environmental service functions of alternative land uses. It was rapidly recognised that AFS have many advantages over monocultures in terms of the increasing demand for multifunctional agriculture and that AFS provide important environmental services. Other recognised potentials of AFS include aesthetic values, buffering of protected areas and agroecotourism.

AFS and Soil Erosion Reduction

The concepts of soil amelioration by trees in AFS have been reviewed by Young and

Buresh and Tian among other authors. Soil improvement in AFS is linked to the growth of nitrogen-fixing trees or deep-rooted trees and shrubs that increase nitrogen availability through biological fixation, recycle plant nutrients from depth and build up soil organic matter.

Formal AFS research initially focused on ways of maintaining soil fertility in annual cropping systems by using leguminous shrub species; e.g. in parkland AFS, in alley cropping and tree-improved fallows. Less research has been carried out on "barrier" AFS, though the use of strips of grass and other annual species to trap sediments and nutrients, slow runoff and increase infiltration has been widely promoted by non-governmental organisations in Central America and Asia. Although many of these AFS studies gave promising results on-station or in researcher managed on-farm trials, for productivity and soil fertility parameters, adoption of alley-cropping systems was disappointing because of: high labour and land requirements; in some cases because of the lack of commercial or home use products from the tree/shrub component; and the long time required to show positive changes.

Planted tree fallows are a potential solution to declining soil fertility due to shortened fallow periods in areas where slash-and-burn is still practised. Nitrogen availability, determined by inorganic soil nitrogen or aerobic nitrogen mineralisation at 0 to 20 cm depth, and crop yields can be significantly higher after a rotation of nitrogen-fixing trees than after other tree species or grass fallows. Relative to herbaceous fallows, greater accumulation of organic material and nutrient storage in biomass, increased root density as well as greater vertical extension of tree roots help maintain nutrient stocks by reducing leaching losses or by taking up nutrients from deep layers. Szott and Palm reported that, in comparison to leguminous herbaceous fallows, leguminous tree fallows greatly increased the total of phosphorus, potassium, calcium and magnesium stocks in the biomass, litter and exchangeable cations/ available phosphorus (soil; 0-45cm). These authors suggested that fast-growing leguminous trees can accelerate restoration of nitrogen, phosphorus and potassium stocks in the crop layer but may not completely restore calcium and magnesium stocks.

The benefits of perennial crop (e.g. coffee and cacao) shade trees include reduced soil erosion as natural litter fall or pruning residues cover the soil and reduce the impact of raindrops, improve soil structure, increase soil nitrogen content and enhance nutrient retention.

Although economic analyses of all the above-mentioned systems are available they do not take into account all the short- and long-term benefits of including the trees, such as improvements or maintenance of soil fertility, nor the possible impact on profitability of service function incentives.

AFS: Contribution to Water Quality

The potential of AFS to help secure water supplies is the least studied service function. The trees in AFS influence water cycling by increasing rain and cloud interception, transpiration and retention of water in the soil, reducing runoff and increasing infiltration. For example, Bharati *et al.* reported that infiltration in areas cultivated with maize or soya, or under pastures, was five times less than under riparian strips cultivated with a variety of plant and tree species, suggesting that the latter had a much higher potential to prevent surface runoff reaching water courses. Moreover, trees in AFS can cycle nutrients in a conservative manner preventing their loss through nutrient leaching. Hence AFS can reduce ground water contamination by nitrate and other substances that are harmful to the environment and human health. As a result of less runoff and leaching, micro-watersheds with forest cover or AFS that cover a high percentage of the soil surface produce high quality water.

A series of studies in Costa Rica have illustrated some of these interactions. For example, rain interception was 16 and 7.5% in coffee (*Coffea arabica*) plantations associated with regularly pruned *Erythrina poeppigiana* (555 trees/ha) or unpruned *Cordia alliodora* (135 trees/ha), respectively. Nitrate losses through leaching were higher from unshaded coffee plantations than from those containing shade trees in areas where high coffee yields had been achieved through large additions of nitrogen from chemical fertilisers probably because of higher rates of transpiration in the AFS. In this country, legislation recognises the environmental services of AFS as well as those of forested land but once again economic analyses that take into account the medium to long term environmental benefits, are needed to determine the true value of the AFS.

AFS: Reducing Emissions of Green House Gases

Highly productive AFS, including silvopastoral systems, can play an important role in carbon sequestration in soils and in the woody biomass. For example, in Latin America, traditional cattle management involves grass monocultures which degrade about five years after establishment, releasing significant amounts of carbon to the atmosphere. Veldkamp estimated that the cumulative net release of CO_2 from low productivity pastures (*Axonopus compressus*) varied from 31.5 (Humitropept soil) to 60.5 Mg C/ha (Hapludand) in the first 20 years after forest clearing. Well managed silvopastoral systems can improve overall productivity, while sequestering carbon, a potential additional economic benefit for livestock farmers. Total carbon in silvopastoral systems varied between 68-204 t/ha, with most carbon stored in the soil, while annual carbon increments varied between 1.8 to 5.2 t/ha.

The amount of carbon fixed in silvopastoral systems is affected by the tree/shrub species, density and spatial distribution of trees, and shade tolerance of herbaceous species. On the slopes of the Ecuadoran Andes, total soil carbon increased from 7.9% under open *Setaria sphacelata* pasture to 11.4% beneath the canopies of *Inga* sp. but no differences were observed under *Psidium guajava*. Soils under *Inga* contained an additional 20 Mg C/ha in the upper 15 cm compared to open pasture.

Few studies have been conducted to determine how payments for carbon sequestration will affect farm income and land-use changes on livestock farms. An *ex ante* analysis showed that farmers can increase income by more than 10% when 20% of grass monoculture pastures are transformed into silvopastoral systems and secondary forest. This economic analysis, conducted on dual purpose cattle farms, suggested that gross potential income generated from carbon stored in the trunks of trees was US$253 per year for a 70 ha farm. Incentives for farmers to adopt silvopastoral systems that store more carbon and prevent pasture degradation are being developed and tested in Colombia, Costa Rica and Nicaragua but a lot more work is needed to realise the full potential of this approach.

AFS: Contribution to Maintenance of Biodiversity

AFS also can play an important role in the conservation of biodiversity within deforested, fragmented landscapes by providing habitats and resources for plant and animal species, maintaining landscape connectivity, making the landscape less harsh for forest-dwelling species by reducing the frequency and intensity of fires, potentially decreasing edge effects on remaining forest fragments and providing buffer zones to protected areas. AFS cannot provide the same niches and habitats as the original forests and should never be promoted as a conservation tool at the expense of natural forest conservation. However they do offer an important complementary tool for conservation and should be considered in landscape-wide conservation efforts that both protect remaining forest fragments and promote the maintenance of on-farm tree cover in areas surrounding the protected areas or connecting them, e.g. in the Central American Biological Corridor.

The degree to which AFS can serve conservation efforts depends on a variety of factors, including the design and origin of the AFS, its permanency in the landscape, its location relative to remaining natural habitat and the degree of connectivity within the habitat, as well as its management and use, particularly pollarding, use of herbicides or pesticides, harvesting of timber and non-timber products and incorporation of cattle, goats, etc. In general, the more diverse the AFS, the lower its management intensity and the nearer it is to intact habitat, the greater its ability to conserve native plant and animal species. Certain AFS, which closely mimic natural ecosystems, provide a variety of niches and resources that support a high diversity of plant and animals, though usually less than that of intact forest. However, even AFS with low tree densities and low species diversity may help in maintaining biotic connectivity.

Equally important is the attitude of local people towards biodiversity conservation and the perceived resulting benefits and losses, which in turn cause local people to favour or discourage native plants and animals. When hunting intensity is high, populations of game species within AFS are unlikely to be viable regardless of whether there is appropriate habitat available.

While there is a growing literature on biodiversity within AFS, important questions still remain about the long-term viability of animal and plant populations in AFS and what will happen to these populations if the surrounding landscape is increasingly deforested. Most studies to date have monitored or inventoried biodiversity within landscapes that still retain some forest cover, have focused on a few taxa and have been conducted on small spatial and temporal scales. Multi-taxa, multi-scale and long-term studies are needed before the true value of AFS for conservation is known.

Despite these limitations in our current knowledge, there is already sufficient evidence that AFS offer more hope for conservation of plant and animal species than the monoculture crops they usually replace. This finding has led to exciting new initiatives to use AFS as tools for conservation in already deforested and fragmented landscapes. Many of these initiatives include either the direct payment to farmers for biodiversity conservation or the certification of products stemming from these AFS as biodiversity or ecologically friendly.

REFERENCES

Anderson L.S. & Sinclair, F.L. 1993. Ecological interactions in agroforestry systems. *Agroforestry Abstracts* 54: 57-91.

Babbar, L. & Zak, D. R. 1995. Nitrogen loss from coffee agroecosystems in Costa Rica: leaching and denitrification in the presence and absence of shade trees. *J. of Environ. Quality*, 24 (2): 227-233.

Beer, J. 1988. Litter production and nutrient cycling in coffee (*Coffea arabica*) or cacao (*Theobroma cacao*) plantations with shade trees. *Agroforestry Systems*, 7: 103-114.

Beer, J., Muschler, R., Kass, D. & Somarriba, E. 1998. Shade management in coffee and cacao plantations. *Agroforestry Systems*, 38: 139-164.

Carter, J. 1995. *Alley farming: have resource-poor farmers benefited*. ODI-Natural Resource Perspectives 3. London UK.

De las Salas, G. ed. 1979. Agroforesty systems in Latin America (Proceedings). CATIE, Turrialba, Costa Rica, 220 pp.

Harvey, C. A., Tucker, N. & Estrada, A. (in press). Live fences, isolated trees and windbreaks: tools for conserving biodiversity in fragmented tropical landscapes? In: G.Schroth, G.A.B. da Fonseca, C.A. Harvey, H. L. Vasconcelos and A.M. Izac, eds. *Agroforestry and biodiversity conservation in tropical landscapes.* Island Press.

Imbach, A.C., Fassbender, H.W., Borel, R., Beer, J. & Bonnemann, A. 1989. Modeling agroforestry systems of cacao (*Theobroma cacao*) with laurel (*Cordia alliodora*) and *Erythrina poeppigiana* in Costa Rica. Water balances, nutrient inputs and leaching. *Agroforestry Systems,* 8: 267-287.

6

Forest Plantations

The question as to what role forest plantations might play in meeting future wood demands is inextricably linked to past, current and future patterns in forest plantation establishment. Indeed, the area of trees already in the ground will determine production in the immediate future. Future planting is dependent on available resources, perceived rates of return, successes of previous planting programmes and perceptions of future supply-demand imbalances.

Unfortunately, however, as already noted, comprehensive forest inventory information about national forest plantation resources is generally scarce and virtually non-existent at the global level. There are several reasons for this, not least the difficulties in differentiating between natural forests and forest plantations. As Solberg *et al* note:

> "Estimating the area of forest plantations presents some challenges. The term "plantation" has varied meanings, and even where a precise definition is available, it is not universally applicable."

A forest plantation will generally be defined according to the extent of human intervention in the forest's establishment and/or management. In many instances, because there is an extensive range of silvicultural practices applied in intensive forest management, the difference between a semi-natural forest and forest plantation is essentially arbitrary. To coin a phrase, a forest plantation is in the eye of the classifier.

Within the framework of the FRA 2000 programme, forest plantations in the tropical and sub-tropical regions are defined as:

Forest stands established by planting or/and seeding in the process of afforestation or reforestation. They are either:

- of introduced species (all planted stands), or
- intensively managed stands of indigenous species, which meet all the following criteria: one or two species at planting, even age class, regular spacing.

For the countries already examined in the temperate and boreal forest component of FRA 2000, the following specification (related to the intensity of management) has been added to the above definition:

> Excludes: stands which were established as plantations but which have been without intensive management for a significant period of time. These should be considered semi-natural.

The above definitions include several ambiguities. For example, how should "regular spacing" be defined, particularly when a stand has been established by broadcast seeding? Similarly, the definitions of "intensively managed" and "a significant period of time" are open to interpretation and the area dimension for the even age class requirement is not specified.

In general, a degree of ambiguity is necessary to ensure that the definition can be applied to a comprehensive range of forests across the globe and to ensure that the definition remains stable through time. There is, however, a necessary trade-off between specificity and robustness. Thus, under these definitions, a number of significant forestry countries including, for example, Finland, Germany and Canada, report having no forest plantations. Conversely, neighbouring countries with seemingly similar forestry practices and philosophies report significant plantation areas.

In the final analysis forest management and silvicultural systems exist on a continuum with "natural" and "artificial" forests occupying different, but perceptually overlapping, portions of the spectrum. In many respects then, especially in the temperate and boreal forest zones, attempting to demarcate between natural forests and forest plantations is an inexact and equivocal exercise.

GLOBAL FOREST PLANTATION AREA

The 1990 Global Forest Resource Assessment estimated that the world's total forest area in 1990 was 3,511 million hectares. This figure was updated in the 1997 State of the World's Forests to give an estimate for 1995 of 3,454 million hectares. However, forest plantations account for only a very small proportion of the global forest area. The most recent estimate of the global area of forest plantations in 1995 is 123.7 million hectares, or approximately 3.6 percent of the estimated global forest area in 1995 for tropical and subtropical countries; and UN for temperate and boreal countries.

The overall concentration of forest plantation resources in a handful of countries is further demonstrated by the fact that only an additional 13 countries have an area of forest plantations exceeding one million hectares. Thus, 18 countries account for 87% of the world's forest plantations. Of the countries with less than 10 million hectares of forest plantations, the largest forest plantation resources are in the Ukraine (4.4 million ha), Brazil (4.2 million ha) and Indonesia (3.0 million ha).

Table 1 provides aggregated estimates of industrial and non-industrial plantation areas in 1995 and includes statistics for the countries with the largest forest plantation areas.

Table 1. The distribution of forest plantations by main countries and region in 1995

Country or region	*Forest plantation area (in million ha)*		
	Industrial	*Non-industrial*	*Total*
North and Central America	18.9	0.3	19.2
United States	18.4	0.0	18.4
South America	5.4	2.8	8.2
Asia	41.8	15.1	56.9
China	17.5	3.9	21.4
India	4.1	8.3	12.4
Japan	10.7	0.0	10.7
Oceania	2.7	<0.1	2.7
Africa	3.6	2.2	5.7
Europe	8.7	0.0	8.7
Countries of the former USSR	22.2	0.0	22.2
Russian Federation	17.1	0.0	17.1
Total	103.3	20.4	123.7

SPECIES USED IN FOREST PLANTATIONS

Several countries with major forest plantation resources stretch across both temperate and tropical zones. Difficulties in clearly distinguishing between tropical and temperate forest plantations in these countries means that the countries have been classified as belonging wholly to one group or the other, usually (but not exclusively) depending on whether the country is classified as developed or developing. This results in a slight bias toward putting those countries that bridge the tropical and temperate divide into the tropical and subtropical category.

At the broadest level, global plantations can be separated into tropical and subtropical plantations and temperate and boreal plantations. These areas can then be further subdivided into forest plantations of hardwoods and softwoods. Tropical and subtropical forest plantation resources in 1995 are estimated to cover 55.4 million hectares, 44.7 percent of the global resource. Hardwood species are estimated to cover 31.4 million hectares, about 56.7 percent of tropical and subtropical plantation area. Softwood species cover 24 million hectares. Plantation forests in temperate and boreal countries are estimated to cover 68.3 million hectares. Softwood species dominate temperate and boreal plantation forests and are estimated to cover 60.7 million hectares or 88.9 percent of the temperate and boreal plantation resource. Temperate and boreal hardwood plantations are estimated to cover 7.6 million hectares.

TROPICAL AND SUBTROPICAL PLANTATIONS

Tropical plantations are dominated by two genera: *Eucalyptus* and *Pinus*. Collectively, these two species groups account for 43.4 percent of tropical plantation areas. Other important species include *Acacia* species, *Tectona grandis*, and Gmelina arborea. Pandey notes, however, that there is a broad range of species utilised in tropical and subtropical plantations. Table 2 gives an estimated breakdown of tropical plantation areas by species group in 1995. The significance of minor species in tropical and subtropical plantations is demonstrated by the area of species classified as "Other softwoods" or "Other hardwoods", which account for 41.6 percent of the total. Around 6 million hectares of "Other softwoods" comprise plantations of Cunninghamia lanceolata in China. This species is not planted on a large scale elsewhere, but the large area in China makes it one of the world's most extensively planted species. Leucaena species and similar multi-purpose and fodder species comprise a significant proportion of "Other hardwoods".

Table 2. Net area of tropical and subtropical forest plantations by species in 1995

Species	*Percentage of total area*	*Area (in ha)*
Acacia auriculiformis	1.4	757,655
Acacia mangium	0.8	454,370
Acacia mearnsii	0.6	325,292
Other acacias	4.2	2,366,990
Casuarina species	1.4	787,200
Dalbergia sissoo	1.1	626,020
Eucalyptus species	17.7	9,949,588
Gmelina arborea	0.7	418,050
Swietenia macrophylla	0.3	151,214
Terminalia species	0.5	303,957
Tectona grandis	4.0	2,246,559
Other hardwoods	24.7	13,920,826
Fast growing pines	10.5	5,923,754
Other pines	15.3	8,614,480
Other softwoods	16.8	9,479,495
Total	100.0	56,325,450

Eucalyptus species are planted extensively throughout the tropics and particularly in subtropical regions. The countries with the largest *Eucalyptus* plantation resources are: India (3.1 million hectares); Brazil (2.7 million hectares); South Africa (557,000 hectares); and Vietnam (479,000 hectares), which collectively account for 69 percent of the total *Eucalyptus* plantation resource.

Fast-growing pine species, particularly *Pinus caribaea,* are also widely planted. The most extensive *Pinus* plantations are, however, in the temperate regions of countries included in the tropical and subtropical classification. Forest plantations in Chile (1.4 million hectares), Australia (833,000 hectares) and South Africa (757,000 hectares), are dominated by *Pinus radiata.* These countries, along with Brazil (1.1 million hectares), have the largest pine plantations in countries classified as tropical or subtropical. *Pinus patula* occupies more than 1 million hectares in Southeast Africa. Slower growing pines occupy around 6.4 million hectares in China.

Other species are less widely distributed. Acacias are planted mainly in Africa, Indonesia and on the Indian subcontinent. Teak (*Tectona grandis*) is predominantly grown in India, Southeast Asia, and parts of Central America and the Caribbean, while Indonesia and Fiji have the largest introduced mahogany (*Swietenia macrophylla*) plantations. Comprehensive information about the main species used in plantations in China (which accounts for 38 percent of the tropical and subtropical plantation area) is not available. However, the most widely planted species in China have been *Cunninghumia lanceolata* (as discussed above), *Eucalyptus* species, *Populus* species and *Pinus elliottii.*

TEMPERATE AND BOREAL FOREST PLANTATIONS

Plantation areas in temperate and boreal countries are generally less easily defined than in tropical countries. For example, Pandey notes:

> Forestry plantations in the developed/industrialised countries are quite different from most of the developing countries. Except for Australia, New Zealand, Portugal, Spain and UK where exotics (eucalypti and/or pines) dominate the plantations, in all other cases, species indigenous to the countries are used mostly in the plantations. The results of the plantations in these countries are often not much different from the results of active natural regeneration. After about 20% of the rotation period, the difference between the planted and naturally regenerated forests almost disappears and often it becomes difficult to assess the actual area under forest plantations.

The UNECE-FAO temperate and boreal forest component of FRA 2000 classifies forests as being natural, semi-natural or plantations. In many instances, forests established as plantations are now classified as being semi-natural, and it is often unclear exactly where, or how, demarcations have been drawn. Because of difficulties in differentiating between natural and plantation forests, particularly in European countries, the temperate and boreal species distribution presented here is only indicative.

Compared with tropical and subtropical forest plantations, softwood species comprise a greater proportion of temperate and boreal plantations. The most important species fall into the Spruce-Pine-Fir (SPF) category with 66.9 percent of temperate plantations comprising SPF species. Pines are, by far, the most common plantation species, constituting around 54 percent of the temperate and boreal plantation resource. Pines are widely distributed throughout temperate regions with the largest resources in the United States of America (almost 17 million hectares, mainly the south) and countries of the former-USSR (11.5 million hectares). A number of other countries have, however, significant pine plantations. Spain, New Zealand, Japan and the Republic of Korea have each planted more than one million hectares of pines in forest plantations.

Spruce and fir species are also mainly planted in the Russian Federation and the United States of America. European countries, most importantly the United Kingdom and Ireland, also have significant spruce resources. Japan has the largest plantation resources of cypress, cedar and larch species, plus significant areas of indigenous sugi (*Cryptomeria japonica*) and hinoki (*Chamaecyparis obtusa*). The most important temperate and boreal hardwood species groups are Quercus and Fagus species. These genera are estimated to account for 6.8 percent of temperate and boreal plantations. Table 3 presents a species breakdown for temperate and boreal forest plantations in 1995.

Table 3. Areas of temperate and boreal forest plantations by species in 1995

Species	*Percentage of total area*	*Area (in ha)*
Pinus species (pines)	54.3%	37,068,804
Picea and *Abies* species (spruce/fir)	12.6%	8,632,269
Larix species (larch)	3.9%	2,644,438
Cupressus and *Chamaecyparis* species (cypress)	3.5%	2,375,260
Cedrus and *Cryptomeria* species (cedar)	7.8%	5,355,310
Other softwood species	6.8%	4,659,592
Eucalyptus species	0.6%	382,228
Quercus species (oak)	5.6%	3,839,151
Fagus species (beech)	1.2%	820,357
Betula species (birch)	0.3%	238,230
Other hardwoods	3.4%	2,297,497
Total	100.0%	68,313,135

TRENDS IN THE AREA OF FOREST PLANTATIONS

Changes in the area of forest plantations vary between years and between countries. This variation is driven by a number of factors, including: government finances; general economic conditions; incentives offered to the private-sector to establish forest plantations; perceptions of the profitability of forest plantation investments; and levels of promotional activities.

In the long-run it would be expected that the absolute area of new planting will decline as countries and investors reach a saturation point. In the short-run however, changes in planting rates may occur as a result of structural changes in the economic environment or investor perceptions. For example, in New Zealand, annual new planting for the period 1993-1997 averaged 75,000 hectares compared with 23,000 hectares in the previous five-year period. This occurred largely because of changes in the perceived value of forest plantations. Similarly, changes in social or environmental factors may generate short-run changes in forest plantation establishment activities or may move the rate of new planting onto a different trend line. For example, if forest plantations become a viable option for carbon sequestration projects, there may be an upward shift in annual rates of new planting in a number of countries.

HISTORICAL PLANTING RATES

Historical estimates of the global forest plantation area are not generally comparable with the current analysis, due to difficulties in defining and classifying forest plantation resources and the generally weak data on areas of new planting, restocking and mortality. As noted earlier, the definition of a forest plantation is problematic in developed countries, while data discrepancies tend to distort statistics for developing countries. Consequently, even recent estimates of the global forest plantation area vary markedly and there are significant discrepancies in reported areas for a number of countries.

In terms of earlier estimates of forest plantation areas, a study by Lanly and Clement, which estimated trends in industrial forest plantation areas in tropical countries to the year 2000, is of particular interest. Lanly and Clement developed "baseline" estimates of forest plantation areas for 1970 and 1975 and produced projections from these of forest plantation area by five-yearly intervals through to 2000. Comparing the estimated industrial forest plantation area in 1995 with their projections (for the same set of countries), their projections appear to have been remarkably accurate. Therefore, it seems reasonable to assume that, for these countries, forest plantation development since 1970 has roughly followed their projections.

Japan and Korea for example, are both countries with relatively high population densities, which have replenished their diminished forest estates. In both countries, rates of plantation establishment have declined quite rapidly as national targets have been achieved. Conversely, forest plantation establishment in Myanmar has accelerated significantly since 1980, with the implementation of a large-scale centralised forest planting programme. Rates of planting in Australia, Chile and New Zealand have fluctuated according to various institutional, legislative and economic changes. Planting rates in Australia show a modest downward drift, while interestingly, planting rates in New Zealand and Chile follow a similar pattern of peaks and troughs.

Planting Rates in Tropical and Subtropical Zone

Pandey reports annual rates of plantation establishment in tropical and subtropical countries of slightly more than 4 million hectares in 1995. About 1.7 million hectares of this is in the tropics and 2.4 million hectares occurs in the subtropics. However, some of this planting is replanting of harvested or failed areas. Pandey notes a general decline in the annual rate of planting in most tropical countries compared with the 1990 Forest Resource Assessment.

Planting Rates in Temperate and Boreal Zone

Aggregated global statistics on historical trends plantation establishment in the temperate and boreal zone are not available, particularly because of difficulties with the definition of forest plantations in North America, Europe and countries of the former-USSR. Nonetheless, it is fair to say that, with a few exceptions, there is a longer history of the establishment and management of forest plantations in the temperate and boreal zone than in the tropical and subtropical zone. In contrast to this, several European countries, such as the United Kingdom and Germany, have two centuries or more of plantation experience. The United States of America, New Zealand, Australia and South Africa all established significant areas of forest plantations before 1930. Japan commenced the bulk of its reforestation programme in 1946 and several North African countries commenced planting around the same time. The Republic of Korea started establishing large areas of forest plantations in 1962. The main exceptions in the temperate and boreal zone are the countries of the former-USSR, where the majority of forest plantations appear to have been established since 1970.

AGE-CLASS STRUCTURE OF FOREST PLANTATIONS

Because of the lack of aggregate national forest plantation inventories for most countries, it is very difficult to compile detailed information about forest plantation age-classes at a global or even regional level. Consequently, FAO has not published a comprehensive age-class analysis since the 1980 Tropical Forest Resource Assessment. However, a large amount of uncoordinated information about the age of forest plantations exists in various forms across countries and regions. This information is very variable in terms of its reliability and how recently it has been produced.

Industrial Forest Plantations

Industrial forest plantations are defined as plantations with the primary purpose of supplying industrial roundwood for the production of sawnwood, wood based panels and wood pulp. Assuming that all of the forest plantations in Europe and countries of the former USSR are industrial forest plantations, the estimated global area of industrial forest plantations in 1995 would be 103.3 million hectares or 84 percent of the total forest plantation area.

Non-industrial Forest Plantations

Non-industrial forest plantations include forest plantations established to produce wood for fuel, plus forest plantations established for reasons other than wood production. Such reasons include: soil and water protection; recreational purposes; and the production of non-wood forest products. The classification used in this analysis excludes plantations of agricultural tree crops, fruit orchards and "non-forest species", such as oil palm, rubberwood and coconut.

Almost all non-industrial forest plantations are likely to be cut at some point, regardless of their original main purpose. The roundwood from such areas may be utilised for wood fuel or for industrial purposes. In some instances, forest plantations that were originally planted to produce wood for fuel are now being used to produce industrial roundwood. One such case is the Republic of Korea where, in the 1960s, more than a million hectares of forest plantations were established to supply wood for fuel. Today, however, wood fuel consumption in Republic of Korea is relatively low and these forest plantations are mainly being used for the production of pulpwood.

Following the assumption that all forest plantations in Europe and countries of the former USSR are industrial forest plantations, the total area of non-industrial forest plantations in the other five geographical regions completes the global picture of the forest plantation resource. The total estimated area of non-industrial forest plantations in 1995 is 20.4 million hectares, accounting for 16.6 percent of the total forest plantation area.

Non-industrial forest plantations in Asia account for 74 percent of the total global area of non-industrial forest plantations. South America and Africa account for the majority of the remainder, with a 14 percent and 10 percent share of the global total respectively. The proportion of non-industrial forest plantations that are less than 15 years is just under 84 percent of the total.

There is greater uncertainty about the age-structure of non-industrial forest plantations given above, compared with the estimates given earlier for industrial forest plantations. Practices such as coppicing and the production of roundwood on a continuous thinning cycle make it difficult to estimate restocking from harvesting figures and may lead to inaccuracies in the estimation of age-classes. Such practices are probably more common in forest plantations grown for wood fuel and for non-productive reasons, than in industrial forest plantations.

FOREST PLANTATIONS IN ASIA

Asia has by far the largest industrial forest plantation resource in the world. However, the bulk of the resource is concentrated in a handful of countries. The three countries with the largest industrial forest plantation resources - China, India and Japan - account for 77 percent of the total area of industrial forest plantations in the region.

It is also interesting to note that the composition of forest plantations in each of these three countries is very different. For example, Japan places a great deal of emphasis on the protection functions of forests. However, this does not generally exclude the production of roundwood in Japan, so all of the forest plantation area in Japan has been classified as industrial forest plantation in this analysis. In contrast, two-thirds of the total forest plantation area in India has been planted for reasons other than industrial roundwood production and this area has been classified as non-industrial forest plantation in this study.

In Japan, around 45 percent of the total forest area is classified as forest plantation and almost all of this area was planted in the period of post-war reconstruction. The main species in these forest plantations are Sugi (*Cryptomeria japonica*), Hinoki (*Chamaecyparis obtusa*), pine and Japanese larch (*Larix leptolepis*). A significant proportion of this forest plantation area is mature or close to maturity (e.g., 54 percent of the area is over 30 years old). However, as Ishihara notes:

> (In Japan)... during the last three decades, the price of wood has been unable to keep pace with the costs of forestry activities. For example, in the past 15 years, plantation costs have more than doubled while stumpage prices for typical Japanese coniferous "Sugi" declined by almost half....At the current level of wood prices, it is estimated that 35.4% of private forests (6.1 million ha.) and 53.7% of national forests (4.2 million ha.) can be profitably harvested.

The establishment of forest plantations in China has accelerated rapidly since 1980 in response to a Central Committee Directive on Vigorously Carrying Out Tree-Planting and Afforestation. Almost all of the area of industrial forest plantations in China has been planted since this Directive was issued. For example, Shi *et al* note that:

> between 1988 and 1992, 16.17 million hectares of timber plantations were established, of which 2.5 million hectares were fast growing and high yield timber plantations.

More than 80 percent of the forest plantation area in China is planted with species that can be used by industry. *Cunninghamia lanceolata* is the predominant species, along with a variety of pine species.

India, the country with the second largest area of forest plantations in the Asia region, has very different objectives for the establishment of forest plantations. More than two-thirds of the forest plantation area in India is classified as non-industrial forest plantation and much of this area has been planted for the production of wood fuel. Not surprisingly, fast growing hardwoods, in particular *Acacia* and *Eucalyptus* species, dominate forest plantations in India. Teak (*Tectona grandis*) is the most important species planted in industrial forest plantations in India, covering a total area of around 1 million hectares.

As in India, a high proportion of forest plantations in Pakistan and Bangladesh has been designated for wood fuel production. Pakistan is also similar to India in that a high proportion of forest plantations have been planted with *Acacia* and *Eucalyptus* species, but a significant proportion of the forest plantation area has also been planted with *Dalbergia sissoo*. Forest plantations in Bangladesh are dominated by species that grow in mangroves, but Bangladesh also has around 70,000 hectares of Teak in forest plantations.

Other Asian countries with over one million hectares of forest plantations include: Indonesia; Democratic People's Republic of Korea; Republic of Korea; Turkey; and Vietnam.

Indonesia has around 3 million hectares of forest plantations, most of which are industrial forest plantations. A considerable range of species has been planted in Indonesia, the most common of which are: *Tectona grandis; Acacia mangium;* and *Pinus merkusii.*

The Democratic People's Republic of Korea has established 2.2 million hectares of forest plantations, with *Larix leptolepis* and *Pinus koriaensis* accounting for around 60 percent of the resource. The Republic of Korea has planted slightly more than 2 million hectares with *Larix leptolepis* and *Pinus koriaensis* also dominating the resource, although a considerable area has also been planted with *populus* species.

The forest plantation resource in Turkey amounts to 1.9 million hectares, comprised mainly of pine species. The most significant species planted in Turkey are Calabrian pine (*Pinus brutia*) and Stone pine (*Pinus pinea*).

Vietnam has established 1.05 million hectares of forest plantations containing a variety of species, of which *Pinus* and *Eucalyptus* species are the most common.

FOREST PLANTATIONS IN AFRICA

The majority of forest plantations in Africa have been established in South Africa (1.4 million hectares) and in the Mediterranean countries of North Africa. In North Africa the countries with the largest area of forest plantations are: Algeria (0.6 million hectares); Morocco (0.6 million hectares); Tunisia (0.3 million hectares); and Libya (0.2 million hectares). Collectively these countries account for 55 percent of all forest plantations in

Africa. Forest plantations are, nonetheless, widely distributed amongst other countries in Africa, with another 16 countries having more than 0.1 million hectares of plantations.

South Africa's plantations largely comprise *Pinus, Eucalyptus* and *Acacia* species (particularly: *Pinus patula; Pinus elliottii; Pinus radiata; Eucalyptus grandis;* and *Acacia mearnsii*). Other countries in Southern Africa (including: Swaziland; Zimbabwe; and Malawi) have also established significant areas of forest plantations with similar species compositions as South Africa.

The plantations of North Africa tend to be of very slow growing species suited to arid and semi-arid conditions. A considerable amount of planting has been carried out in dune stabilisation projects in an attempt to halt desertification. The Algerian "green dam" project is particularly notable in this regard. Indeed, although very little information is available about this resource, the second largest industrial forest plantation resource in Africa is in Algeria. The predominant species grown in Algeria are, however, very slow growing. The most common species used are Cork oak (*Quercus suber*) and Aleppo pine (*Pinus halepensis*). Other common forest plantation species used in North African countries include: *Pinus halepensis; Pinus brutia; Eucalyptus camaldulensis; Eucalyptus globulus; Eucalyptus gomphocephala;* and many species of *Acacia*.

Africa has the highest proportion (36 percent) of non-industrial plantations to total plantations of all the geographical regions and probably the highest proportion of forest plantations used for wood fuel production. In particular, Sudan, Ethiopia and Rwanda, are countries with relatively large areas of forest plantations used for wood fuel, which are mostly planted with *Eucalyptus* or *Acacia* species.

Several special purpose species are also grown extensively in African plantations. Cork oak (*Quercus suber*) is widely planted in Algeria; *Acacia senegal* (a source of gum arabic) is grown in plantations in Sudan, Senegal and several other Sahelian countries; and *Acacia mearnsii* (wattle) is grown for its bark in South Africa, Zimbabwe and Swaziland.

The structure of age-classes in industrial forest plantations in Africa contains slightly larger areas in younger age classes, reflecting increases in planting rates in recent years. However, the proportion of younger industrial forest plantations is not as great as in Asia. In the older age classes, South Africa accounts for the largest share of industrial forest plantations. In fact, 79 percent of all industrial forest plantations over 30 years old are in South Africa.

FOREST PLANTATIONS IN OCEANIA

The most comprehensive information about forest plantations at a regional level can be found in Oceania. Australia and New Zealand account for 95 percent of forest plantations in the region, having areas of 1 million hectares and 1.5 million hectares respectively. Both of these countries have recently produced comprehensive inventories of their forest plantations, giving numerous details of their plantation forest resources. The other country in the region with a significant area of forest plantations is Fiji, with 0.1 million hectares of forest plantations.

The main species used in forest plantations in Oceania is *Pinus radiata*, which accounts for 91 percent of the forest plantation area in New Zealand and 62 percent in Australia. Other pine species, most notably *Pinus caribaea* in Fiji and *Pinus caribaea* and *Pinus oocarpa* in Northern Australia, account for most of the rest of the forest plantation area that is planted with softwoods. *Eucalyptus* is the most common hardwood planted in forest plantations in the region and is found mostly in Australia. However, Fiji also has significant areas of forest plantations containing Mahogany (*Swietenia macrophylla*) and Teak (*Tectona grandis*).

The age-class structure is more balanced than in most of the other regions, due to an early recognition of the role that forest plantations could make in meeting the region's demand for wood. Both New Zealand and Australia began establishing forest plantations before 1930 and significant areas are currently reaching maturity or are already into second or third rotations. The predominance of areas under 35 years old in the region reflect the rotation ages typically used in forest plantations in the region (i.e. a significant proportion of these areas are replanted areas rather than areas of new planting). New Zealand, Australia and Fiji all anticipate significant increases in wood production from their industrial forest plantations during the next decade.

FOREST PLANTATIONS IN NORTH AND CENTRAL AMERICA

The United States of America accounts for nearly all of the forest plantation area in North and Central America, with a total area of 18.4 million hectares. Among the other countries in the region, only Cuba (with 0.4 million hectares), Mexico (0.2 million hectares) and Costa Rica (0.1 million hectares) have significant areas of forest plantations.

The bulk of forest plantations in the United States of America (90 percent) are located in the Southeast and South-central regions and around 85 percent of forest plantations in the United States of America are planted with pine species. Loblolly pine (*Pinus taeda*), Shortleaf pine (*Pinus echinata*), Longleaf pine (*Pinus palustris*) and Slash pine (*Pinus elliottii*) are the most common species found in these forest plantations.

Cuba, Costa Rica and Mexico have all planted a variety of species in their forest plantations. In Cuba, fast-growing pine species (notably *Pinus caribaea, Pinus tropicalis* and *Pinus cubensis*) account for 48 percent of the forest plantation area. The main species planted in Costa Rica is *Gmelina arborea* (accounting for 34 percent of the forest plantation area), although significant areas have also been planted with Teak (*Tectona grandis*) and alder (*Alnus acuminata*). In Mexico, pines (including: *Pinus patula; Pinus ayacahuite;* and *Pinus strobus var. chiapensis*) account for most of the area of forest plantations planted with softwood species. A wide range of species (including: *Eucalyptus; Acacia;* and *Casuarina* species) can be found in the forest plantations planted with hardwoods.

The vast majority of forest plantations in the United States of America and Costa Rica are classified as industrial forest plantations. In contrast, more than 40 percent of forest plantations in Cuba and more than 60 percent in Mexico are classified as non-industrial forest plantations.

Again, the distribution of areas amongst different age-classes is relatively balanced. In total, nearly all of the forest plantation area in the region is less than 30 years old and some of these areas are second or third rotation forest plantations. For example, much of the forest plantation area in Cuba and Mexico was first planted in the mid to late 1950s, although the majority of these areas are now less than 20 years old. The exception to this is Costa Rica, which reported only 2,700 hectares of forest plantations in 1980 in the 1980 Tropical Forest Resource Assessment. This suggests that the majority of forest plantations in Costa Rica have been established in the past 15 years.

FOREST PLANTATIONS IN SOUTH AMERICA

The total estimated area of forest plantations in South America in 1995 is 8.2 million hectares. The following three countries account for 82 percent of this resource: Brazil (with 4.2 million hectares); Chile (1.7 million hectares); and Argentina (0.8 million hectares). However, despite the dominance of these three countries, large areas of forest plantations can also be found across most of the rest of the region, with 8 of the other 13 countries in South America having more than 0.1 million hectares of forest plantations each.

The most common species used in forest plantations in South America are fast-growing *Pinus* and *Eucalyptus* species. The total area of forest plantations planted with *Eucalyptus* is about 3.9 million hectares, closely followed by forest plantations of *Pinus* species, which account for a further 3.5 million hectares.

Although, for convenience, the whole region has been classified as tropical and subtropical for this study, large areas in the southern part of the region are in the temperate forest zone. The forest plantations in these areas are dominated by pines (most notably *Pinus radiata, Pinus elliottii,* and *Pinus taeda*), which account for 49 percent of the forest plantation area in Argentina and 78 percent in Chile. Pines also account for 80 percent of the tropical forest plantation area in Venezuela, where the most common species used is *Pinus caribaea*.

Eucalyptus species are the most commonly used species in the remainder of the tropical and subtropical zone in South America, accounting for 65 percent of the forest plantation area in Brazil, 90 percent in Peru and 80 percent in Uruguay. Particularly common species include: *Eucalyptus globulus; Eucalyptus grandis; Eucalyptus saligna; Eucalyptus urophylla; Eucalyptus deglupta;* and the F-1 hybrids of the latter two species, which are used in the tropical parts of Brazil.

It is estimated that industrial forest plantations account for 74 percent of the total forest plantation area in South America, while Brazil, Peru, and Uruguay have the largest non-industrial forest plantation resources.

The South America is similar to the other tropical and subtropical regions, in that the age-class distribution is dominated by areas in the younger age-classes, reflecting increases in new planting in recent years. The area of forest plantations under 10 years old accounts for 45 percent of the total industrial forest plantation area.

FOREST PLANTATIONS IN EUROPE

As noted earlier, many countries in Europe have not previously made the distinction in their national forest inventories between forest plantations and other types of forest. For example, countries have reported forest plantation areas as part of the temperate and boreal component of FRA 2000. However, in many cases, this is the first time that this has been attempted. Consequently, it has been very difficult to collect and collate historical statistics on forest plantation areas.

The main difficulty encountered in this analysis has been to match historical information about areas planted and replanted, with the areas of forest plantations reported in the temperate and boreal component of FRA 2000. This is partly because it is suspected that many areas reported as forest plantations in the historical data are now classified as semi-natural under the terms and definitions agreed for FRA 2000. Therefore, it has been necessary to make a number of assumptions in order to resolve discrepancies between the historical planting data and the data reported in FRA 2000. In doing this, it has been assumed that areas recently reported as forest plantations are still reported as forest plantations in FRA 2000, while areas planted some time ago may now be classified as semi-natural forests.

Sweden can be taken as an example of how this has been done. Widespread planting of *Pinus contorta* in Sweden began in the early-1970s and Sweden now reports around 550,00 hectares of forest containing this species. This number is very close to the 572,000 hectares of forest plantations reported by Sweden in the FRA 2000. Consequently, it has been assumed that most forest plantations in Sweden have been planted with *Pinus contorta* since 1970 and that earlier recorded areas of forest plantations are now classified as semi-natural forests.

Several countries in Europe do however, classify a high percentage of their forests as forest plantations in FRA 2000. For example, Ireland and Malta classify 100 percent of their forest area as forest plantations. Other countries reporting a high proportion of forest plantations include: Denmark (92 percent of the total forest area); United Kingdom (57 percent); and Belgium (46 percent). In contrast, several European countries with significant forest resources report no forest plantations at all. These countries include: Austria; Finland; Germany; and the Czech Republic. These countries are not included in this analysis.

Five countries in Europe account for two-thirds of the area of forest plantations in Europe. Spain has the largest forest plantation area with 1.9 million hectares, followed by the United Kingdom (with 1.4 million hectares), Bulgaria and France (with 1.0 million hectares each) and Portugal (0.8 million hectares).

Spruce, pine and fir species account for the largest share of the forest plantation area in Europe. Forest plantations in Spain and Portugal are dominated by pines (especially *Pinus pinaster, Pinus halepensis, Pinus pinea* and *Pinus radiata*) and *Eucalyptus* species. The United Kingdom has a high proportion of forest plantations planted with Sitka spruce (*Picea sitchensis*) and Douglas fir (*Pseudotsuga menziesii*). In France, the most common species used in forest plantations are *Populus* species, Norway spruce (*Picea abies*) and Douglas fir (*Pseudotsuga menziesii).*

The distribution of forest plantation area across age-classes is quite balanced in Europe, with significant proportions of the forest plantation area between 30 and 50 years old and over 50 years old. Of the five countries with the largest forest plantation areas, France has the greatest proportion in older age-classes. Age classes in the United Kingdom are fairly evenly distributed, while forest plantations in Spain, Portugal and Bulgaria tend to have a relatively high proportion of younger aged forest plantations.

FOREST PLANTATIONS IN COUNTRIES OF FORMER USSR

The 15 countries of the former USSR have also been included in the temperate and boreal component of FRA 2000 and all have reported that they have some areas of forest plantations. The Russian Federation accounts for the largest share of forest plantations in this region (with 17.3 million hectares), but only 2 percent of forests in the Russian Federation are classified as forest plantations. The Ukraine accounts for the next largest share in the region, with 47 percent of the total forest area classified as forest plantation. Belarus, Estonia, Latvia and Lithuania have much smaller, though still significant, areas of forest plantations. Although the other countries of the former USSR all report having small areas of forest plantations, very little information is available about forest plantations in these countries, so they have been excluded from the analysis.

Very little information is also available about the species composition of forest plantations in this region. Pandey reported the following species composition for forest plantations in the USSR in 1988: pines - about 52%; spruce - 24%; oak - 6%; and cedar - 1%. It is unlikely that this species distribution has changed significantly since 1988.

Information about the age structure of forest plantations in the Russian Federation is available in the State Forest Account and much of this information has been made accessible through forestry research projects carried-out at the International Institute for Applied Systems Analysis (IIASA) in Laxenburg, Austria. For example, a time-series showing the accumulated area of forest plantations is given in Shvidenko and Nilsson and, from this, an approximate age-class structure for forest plantations in the Russian Federation can be derived. Shvidenko and Nilsson note that, between 1961 and 1993, significant efforts were made to establish forest plantations in the Russian Federation, but also that survival rates were typically only 55 to 60 percent.

A significant proportion (56 percent) of the area of forest plantations in the Russian Federation was planted before 1973 and is consequently now over 25 years old. Afforestation efforts slowed markedly after 1988 and, given the dramatic fall in roundwood production in the Russian Federation that has occurred since 1992, it seems reasonable to assume that the rate of forest plantation establishment during the period 1990 to 1995 has been negligible. A similar pattern in forest plantation establishment can be seen in the Ukraine, where large-scale afforestation and forest regeneration efforts took place throughout the 1960s, but have subsequently been reduced, particularly after 1980.

FOREST PLANTATION YIELDS AND ROTATION LENGTHS

In many countries, forestry policy has encouraged the development of forest plantations for roundwood supply. In some cases, this has been done to meet rising demand where existing supplies from the natural forest are very low (e.g., forest plantations for wood fuel production in India). In others, forest plantations have been seen as an important supplement to supplies from natural forests (e.g. Indonesia) or even as a complete substitute to them (e.g. New Zealand). Given this trend, the following questions are becoming increasingly important for the analysis of such policies:

- what role are forest plantations currently playing in regional and global roundwood supply;
- what is the future potential supply of roundwood from forest plantations;
- what are the ecological impacts of forest plantations and how sustainable is this model of forestry development in the long run; and
- what are the effects of forest plantations on forest values as a whole?

In attempting to shed some light on the former two questions, four variables are of fundamental importance: the current and future likely area of forest plantations; the species used in forest plantations; the age-class distribution in forest plantations; and the growth or mean annual increment that can be expected in forest plantations. A number of additional factors are also important, such as: the rotation lengths and silvicultural regimes used in forest plantations; tree mortality; the potential for genetic improvements to increase yields; and the overall quality of management in forest plantations, but these factors are largely subsidiary to the main variables of forest plantation area and yield.

FOREST PLANTATION YIELD ESTIMATION

In order to model future potential roundwood production from forest plantations, it is necessary to have information about average forest plantation yields that might typically be achieved under normal operational conditions. Unfortunately, such information is both scarce and often imprecise.

A vast amount of literature is available about the yields of different species achieved in forest plantation research trials and experiments. However, it is likely that the yields obtained in forest plantations managed on a commercial scale will vary considerably from these results, due to variations in land quality and the quality of establishment and silviculture. It is also likely that, overall, commercial operations will achieve generally lower yields than those reported in the research literature, because the quality of establishment and silviculture is likely to be lower in commercial operations.

Assessing the yields that might be obtained under normal operating conditions is crucial, because a small variation in yield can have a major impact on final harvest volumes. For example, if research results suggest that a species will grow at 7 m^3/ha/ year, but only 5 m^3/ha/ year is achieved in commercial operations, using the former figure would overestimate final harvesting volume by 40 percent. Thus, at an aggregate level, it is important to avoid overestimating likely potential roundwood yields, which could result in biased forecasts of total future potential roundwood production.

Estimated Current Production Potential

Based on all of the information described above, a simple production forecasting model has been constructed for this exercise. This model produces projections of the volume of roundwood that could be produced from the world's forest plantations based on their area, species, type, yield and age structure. It must be stressed that this is a projection of potential production and that actual production may differ from this for a number of reasons. However, due to the high levels of investment in forest plantations, it seems likely that most forest plantations will be fully utilised for wood production and thus, that actual production will be quite close to potential production.

Potential Roundwood Production

Current statistics collected about roundwood production do not differentiate between roundwood produced from the natural forest and roundwood produced from forest plantations. Thus, the production forecast model was first used to estimate how much industrial roundwood might have been produced in 1995 in industrial forest plantations. Based on the age structure of industrial forest plantations in 1995, the model suggested that industrial forest plantations could have accounted for about 331 million cubic metres of industrial roundwood production or about 22 percent of total global industrial roundwood production.

Assuming that non-industrial forest plantations are used mainly for wood fuel production, the model also suggested that non-industrial forest plantations could have produced about 86 million cubic metres of wood fuel or just over 4 percent of total global wood fuel production.

Estimated Product Mix

Generally, very little information is available about the mixture of roundwood sizes and qualities that might be produced in industrial forest plantations and, consequently, the products that might be produced from such wood. In only a few countries, where forest plantations account for almost all industrial roundwood production, can this information be easily obtained. For other countries, estimates can be made based on the forest plantation species mixture, but these are likely to be highly speculative and potentially misleading.

Only five countries have a proportion of industrial roundwood production from forest plantations that is sufficiently high enough to make a reasonably reliable assessment of the product mixture coming from their forest plantations.

Forest plantations in all five countries are predominantly planted with temperate zone species and, consequently, the product mixtures coming from forest plantations in these countries can not be used as a guide to what might be produced in tropical forest plantations. Furthermore, it is quite likely that a very different product mix might be produced in temperate forest plantations with very different species mixtures to these countries.

Between 30 percent and 50 percent of industrial roundwood in New Zealand, South Africa and Chile is used for the production of wood-based panels, wood pulp, mining and other uses. Assuming that the majority of roundwood exports are exports of sawlogs, this would suggest that pulpwood production from industrial forest plantations is, therefore, between 30 percent and 50 percent, while sawlog production is between 50 percent and 70 percent. Denmark and, in particular, Ireland appear to produce a slightly higher proportions of sawlogs (70 percent to 80 percent). Given that growth rates in the first three (Southern) countries are somewhat higher than in the latter two, this would suggest that, where forest plantation yields are higher, a greater proportion of the roundwood produced from forest plantations might be in the pulpwood category.

Economic factors tend to favour using short rotations in forest plantations which, in turn, tends to favour the production of pulpwood rather than sawlogs. However, as the above differences demonstrate, where forest plantation yields are lower, it appears that forest plantation owners tend to pursue a strategy of producing relatively more higher value sawlogs and veneer logs. This may have something to do with the better structural qualities of roundwood coming from slower growing forest plantations.

REFERENCES

Carle, J., Del Lungo, A. and Vuorinen, P.. 2001. *Preliminary Analysis of Forest Plantation Development 1980-2000*. Proceedings of the ANPA-IUFRO-Dalarna University International Conference on Nursery Production and Stand Establishemnt of Broad-Leaves, Rome, 7-10 May 2001

FAO 1995. *Forest resources assessment 1990, Tropical forest plantation resources.* (FRA 1990) FAO Forestry Paper 128. FAO, Rome, Italy.

FAO 2000. *The global outlook for future wood supplies from forest plantations.* Brown, C., FAO Working Paper GFPOS/WP/03. FAO, Rome, Italy. 129 pp.

Sedjo, R.A. 2001. "The role of forest plantations in the world's future timber supply". *The Forestry Chronicle* Vol 77. No. 2: 221-225.

7

Smallholder Agroforestry Systems

During recent years, environmental agencies have done a valuable job in drawing public attention to the rate of tropical forest destruction and mobilizing awareness of the need for mole effective forest protection policies. During this century the area of tropical forest of the world has declined by more than a half. FAO's latest estimates expect a further 10 to 15 percent decline by the end of the century, and it is possible that, unless something is done to reverse the present trend, by the middle of the next century, the bulk of the tropical forest ecosystem as we know it could disappear. Botanists, ecologists and environmentalists have pointed out the irreversible loss to mankind which would. result, citing, in particular, the loss of genetic material and the potential contribution to human welfare of drugs and medicines available from tropical woody plants. Many international conferences have been held to help create better political awareness of these issues.

However, a deliberate shift in the emphasis of conservation and development strategy is needed. If we are to ensure preservation of a significant part of the world's remaining tropical forest ecosystem, we should focus more on how to improve the incomes and quality of life of the 200 million subsistence farmers living in a state of shifting cultivation in tropical forest areas. Only the briefest glance at the history of agricultural settlement in Europe, North America and elsewhere is needed to suggest that any policy aimed at halting the present process of forest destruction while completely excluding people from the tropical forest areas is unlikely to succeed. Attacking the root cause of forest destruction-rural poverty in forest areas - and providing small farmers with a viable alternative to shifting cultivation are the key issue. An essential first step would be the recognition that a large part of the "forest destruction" taking place in tropical developing countries, which has generated such an emotional response from agencies in predominantly temperate-zone developed countries, is, in fact, a logical shift in land use to more productive agriculture.

What can be done in practical terms to make it possible for small farmers to abandon forest cutting and shifting cultivation, to adopt sustainable farming systems and to become part of more stable rural communities? What are the most appropriate choices and techniques? A few examples of project experiences, some successful and some less so, financed partly by the World Bank may help in the search for solutions for reducing the risk of continued ecological degradation. Two of such agricultural land-settlement projects, both of which involved forest clearing followed by, in the first case, agricultural tree cropping and in the second, livestock development, are examined below.

JENGKA TRIANGLE LAND SETTLEMENT PROJECT, MALAYSIA

The Jengka Triangle in the state of Pahang covers about 120 000 ha, of which about half are considered suitable for agricultural development. The area was identified in the early sixties as favourable for large-scale tree-crop development and settlement. While earlier land settlement took place in smaller schemes scattered throughout Malaysia and close to existing infrastructure, the Jengka Triangle was to be the largest attempt at that date for the development of virgin tropical forest lands.

In 1965, a technical assistance grant was made by the World Bank to the Government of Malaysia to help finance a land-use study of the area and the preparation of a regional development plan. The "master plan" completed in 1967 called for comprehensive development of the Triangle, comprising settlement, in a first phase, of some 9 000 farm families cultivating about 40 000 ha of oil palm and rubber; systematic exploitation of forest resources prior to settlement; urban development, including the establishment of three new townships; and extensive infrastructure development.

A first Jengka Triangle Project, commencing in 1968, planted 12000 ha of oil palm and 1600 ha of rubber. A second project, commencing in 1970, developed a further 7 000 ha of oil palm and 6 000 ha of rubber. Physical works included clearing of forest land, construction of houses, offices and stores to accommodate settler families, and recruitment of management and support staff. A palm oil mill was constructed together with appropriate roads, water systems, and educational, health and other social services. About 300 ha were developed for crop diversification trials on a commercial scale. Each settlement comprised about 4 ha of planted oil palm or rubber, and a house lot of 0.1 ha for growing food crops. A third loan, made in 1973, will complete the programme.

According to the three criteria defined earlier, the project can be judged successful. Rural incomes of the 9 000 families settled in the first phase have shown a four-fold increase. Settler turnover rates are low (two percent) and the village communities are expected to remain stable. By careful forward planning and the carrying out of appropriate land-use and soil-capability surveys prior to settlement, about 80 000 ha of forest, comprising 60 percent of the project area, were excluded from agricultural settlement. Cultivation was confined to the flatter areas and hill slopes, and river banks were retained under forest. The higher levels of rural income and stable communities in the project area have reduced the risk of shifting cultivation and further forest destruction. Also, it seems reasonably certain that the cropping patterns developed in Jengka, based on perennial tree crops, are sustainable, given appropriate fertilizer application. The economic rates of return have been higher than expected, and Malaysia's exports of palm ail have been a very significant source of foreign exchange earnings.

On the negative side, there were several problems. Settling families had difficulty in protecting their crops from wild pig and other animals because of the close inter-relationship of forests and settled lands; attempts to increase revenues from salvage logging in the area prior to settlement, by establishing a sawmill and plywood mill, have not been very successful. Finally, controversy arose over the relatively high cost of the project (US$15 000 per settled family) and the extent to which this type of project is replicable. Lower cost criteria have now been introduced for future World Bank involvement in settlement projects.

To maintain at least part of the remaining tropical forest ecosystem intact, the Malaysian Government in 1976 created an Environmental Ministry and prepared a comprehensive environmental plan for the country, aiming at setting aside more than 1 million ha of forest as permanent biotic reserves and national parks. Of this, 0.5 million ha have already been reserved.

As a model for replication, in other countries, the intensive land-use and soil-capability surveys carried out prior to the Malaysia Jengka project are particularly noteworthy. The perennial agricultural tree crops being grown provide an effective soil protection and catchment area cover, and the prospect of sustainable income for the farmers. Such perennial agricultural tree crops already cover about 25 million ha of the world's former tropical forest lands, market prospects for most of these crops are good and further expansion of something in the order of an additional 2 million ha can be expected between now and the turn of the century.

CAQUETA SETTLEMENT PROJECT, COLOMBIA

In this project, land settlement was spontaneous, less formalized and less successful than in Jengka, and was based mainly on a livestock farming system. Colonization of the tropical forest areas of Colombia started in Caqueta during the rubber boom, earlier in this century. In the late thirties large numbers of settlers 'began to move in as word spread that they could take possession of public land and that the area was very well suited for livestock. Government support started in 1959 with a directed settlement scheme organized by Caja Agraria, which failed because of poor selection of settlers and inadequate supervision of credit beneficiaries.

In 1969, the Government of Colombia requested World Bank assistance in development of a continuation of the settlement programme. A loan of US$8 million was made in 1971 for a first-phase Caqueta Project, which was to be developed over three years and administered by a. new settlement agency, INCORA. It was to benefit the 8 000 settler families living in the area. The first phase provided long-term livestock loans for 4 500 settlers, construction of 380 km of roads, 90 primary schools, six health centres and improvement of INCORA's administration. Settlement costs were estimated at US$20 million.

In practice, the project suffered from a number of problems including considerable price increases in all fields, unexpectedly difficult physical conditions affecting., in particular, the road construction programme, and lack of participation by settlers in constructing schools under self-help programmes. Toward completion of disbursement, the project design was changed and, in 1975, a second loan was made taking into account difficulties encountered under Phase I. It was concluded when defining the second phase that, while it was premature to observe any improvement of beneficiaries' incomes, the possession of a basic livestock herd had enabled participants to maintain themselves on their current holdings (averaging 85 ha) instead of continued dependence on shifting cultivation. By importing 60 percent of the breeding cattle into Caqueta, the project had "markedly improved the development prospects of an area designed to play a major role in Government's efforts to develop livestock production".

Outstanding problems, such as the lack of technical assistance to farmers, inadequacy of road maintenance, and the provision of social services were to be rectified in the second project phase.

The Caqueta project has been controversial. Kirby, for example, has commented in *Pacific Viewpoint:*

> "Not only are most farmers operating a farm unit smaller than that regarded as viable in a beef breeding/ fattening economy, but that the tendency toward a bimodal structure is accentuated by the inability of small farmers to buy cattle. Credit is available for the purchase of foundation stock, and, with an inflation rate of more than 20 percent, credit bears a negative rate of interest of 12 percent per annum after a three-year grace period. But new colonists are very wary about credit for cattle purchase, for, if animals die, or are rustled, the loan must still be repaid. Credits for land clearance or pasture are rarely sought since the value of the improved land will be directly dependent on grazing animals not necessarily available to offset its cost. In addition, the Caja de Crédito Agrario has an understandable tendency to lend money to established farmers, where supervision is easier and repayment guaranteed by the collateral security of an existing herd. In summary, the situation in Amazonia is one of very slow improvement in the lives of the new settlers. In 1971, only 55 percent of Medina's sample, in Caqueta and Putumayo, would have stayed on their farm if the possibility existed of their moving elsewhere. For the majority, life is one of shifting cultivation of subsistence crops, living on informal shopkeeper credit."

It would be premature to draw any firm conclusions about the project's possible long term impact on rural incomes. But this project does highlight the major issue concerning planned settlement in the Latin American tropical forest regions-that of the extremely poor quality of some of the forest soils and the difficulties of ensuring sustained livestock and crop production. Much publicity has been given to the degradation of former tropical forest lands in Brazil, caused, for example, by badly managed livestock schemes. By contrast, Sanchez has presented a body of evidence from trials carried out by the Centro Internacional de Agricultura Tropical, Call, Colombia (CIAT), and other agencies that, given appropriate fertilizer treatment, stocking density and agronomic management, a considerable proportion of the acid latosols of the Amazon region is capable of sustained agricultural crop or livestock production. Several fairly large-scale pilot programmes are under way, the results of which could be highly significant for future development in the Amazon region.

Regarding the extent of adequate provision for protecting the forest resources in this settled area, the Caqueta Project experience was an acknowledged failure. At the outset of the project, a deliberate attempt was made to set aside an area of 20000 ha as a permanent forest reserve, but within a year, despite expenditure on forest guards' housing and protection services, the area was invaded by colonizing families.

To ensure an adequate supply of fuel, building poles and timber for incoming settlers and to maintain the protective role of the forests, the Caqueta Project area which was originally part of an officially declared "Amazon Forest Reserve" was made the target of special resolutions3 aimed at ensuring that colonization should take into account the need to preserve the forests. The law required recipients of more than 50 ha of public lands to keep 20 percent under forest and it allowed the Government to maintain 10 percent of the area as a protective zone. In practice, the farmers' obligation under this law proved impossible to enforce; the experience on fully developed farms showed that, on average, settlers would maintain not more than five percent of their land under forest for the protection of a spring, or for the supply of housing and fencing wood.

This experience suggests the need for greater flexibility in defining forest laws which decree that an arbitrary percentage of settlement areas should be retained as forest cover, a common feature of land-settlement projects. The Caqueta farmers' decision to protect only five percent of the forest land in order to ensure basic needs for fuelwood and other forest products would seem quite rational in the light of experience elsewhere, which suggests that an average rural family might need something between 250 and 500 trees (less than 0.5 ha) to maintain basic domestic needs. The relevant point is that the main beneficiaries of the various government resolutions aimed at protecting a larger area than this would be farmers situated downstream from the Caqueta Project area, who would benefit from protection of the river head-waters, reduced flooding and sedimentation. These "external" benefits have little relevance to farmers within. the Caqueta Project area, and it is hardly surprising that they should regard the 20 percent restriction primarily as an obstacle standing between them and the possibility of increasing family income by developing additional food cropping areas or acquiring more livestock.

The broader issue raised here is whether, in fact, retention of, say, 20 percent forest cover is the only way to ensure effective catchment area protection. While there is plenty of scientific evidence to show that undisturbed natural forests provide an optimum cover for ensuring adequate soil protection and regulating downstream cow, there is also evidence from many parts of the world, including tropical areas, that other forestry, agriculture and livestock farming systems can also provide adequate catchment protection provided care is taken over soil conservation measures, and livestock numbers are maintained in balance with the carrying capacity of the land. Seen in this light, an alternative approach to designing the Caqueta Project might have placed greater emphasis on soil conservation measures and on the backup extension services needed to ensure adequate husbandry practices. For protecting forests on very steep slopes and along river banks in the project area, greater flexibility in selecting areas for protection and closer consultation with incoming settlers on this aspect might have produced different results. Recently, the project's forestry component has been revised along these lines and progress is being monitored to assess the impact of these changes in project design.

A second major issue which arose during the formulation of the Caqueta Project, and which has considerable relevance to settlement schemes in other parts of the tropics, was the question of how to increase returns from logging operations prior to settlement. Before 1975, land-clearing operations in Colombia had resulted in the felling and burning of 500 000 ha of forest. At the time of project preparation, clearing was proceeding at the rate of 30000 ha a year. Every year, it was estimated that 2 million cubic metres of mature timber were being cleared, only one percent of which was sawn and sold, and the rest burned. Of total standing biomass volume of something between 250 and 300 cubic metres per ha, only 25 trees averaging 40 cm diameter or more were suitable for processing into lumber or plywood-and of that, only 17 cubic metres were of species which were marketable. Although a further 30 cubic metres (40 percent) were suitable for charcoal burning and fuelwood production, they could not be used for this purpose because both local and nearest potential export markets were saturated. After intensive study of this issue, it was concluded:

> "It is considered quite impractical, if not impossible, to rationalize felling at the present stage of development of the Caqueta Project. Whatever benefits could be obtained from a rationalized forest exploitation should be weighed against the delays it would cause in developing the area through spontaneous colonization. The studies carried out show that, under these conditions, rationalized forest exploitation would not be economic".

TAUNGYA SYSTEM, KENYA

Kenya's plantation forestry programme has a number of features of general interest including the role which plantations could play in reducing pressure on the natural forest ecosystem; the provision for the setting aside of specified nature reserves; the fact that some of the past shift in land use from forestry to agriculture was based on systematic long term catchment area studies of the likely impact of different cropping patterns on stream flow and downstream agriculture.

Kenya's forests cover about 2.5 million ha (about four percent) of the country's total area (16 percent of the land area receiving more than 850 mm of rainfall). Over the past 50 years, the indigenous forest has been continuously exploited for the production of sawntimber and other forest products. Because natural regeneration of indigenous forest species takes between 60 and 100 years to produce timber of usable size, the Government, over the last 30 years, has been replacing some of these forests with faster growing exotic softwood plantations. To date, a total of about 160000 ha of industrial plantations have been established, representing seven percent of the total forest area.

In 1969, the World Bank made a loan of US$2.6 million to finance part of the costs of a six-year time-slice of this plantation development programme. The aim was to establish 28 000 ha of plantations during this period and it was successful in meeting the target. In 1976, a second loan of US$10 million financed the continuation and expansion to cover the whole of the industrial plantation programme of the Forestry Department. This project is due for completion in 1980 and a third phase will, simultaneously, concern rural afforestation and industrial capacity needs for processing the expanding raw material base.

Most of Kenya's afforestation programme has been carried out using the "taungya" system. In Kenya, forestry workers grow mainly maize, beans or potatoes for a period of four or five years, after which the plantation is grown on as a monoculture forestry crop until ready for harvesting. Pines and Mexican cypress were the main species used.

The forestry plantation programme provides sustained employment for some 5 000 persons. Kenya's forest villages, more than 100 of which have been established over the last 30 years, sustain stable forest communities dependent on a combination of agriculture and forestry work for their livelihood. Many of the forestry workers are second-generation forest villagers. As the forestry programme has proceeded, secondary employment opportunities have been generated in logging, sawmilling, pulp and paper and furniture factories.

The: new forestry plantations have a wood productivity some 15 times greater than that of the indigenous forest which they are replacing. The deep volcanic soils on which the plantations are being established are capable of sustained cropping, although recent research work suggests some fertilizer application may prove necessary between rotations.

Two points of general interest arise from this project experience. The first is the role which such compensatory plantations can play in relieving the pressure on indigenous catchment protection forests. It is from the 2.4 million ha of indigenous forest that most of Kenya's important rivers and streams originate. Prior to the fifties, more than 90 percent of timber production came from these indigenous forests. Timber-concession licences had been allocated under long-term contract arrangements covering most of the accessible forest area.

Today, in 1980, the compensatory plantations which have been established in Kenya -and cover less than 10 percent of the former indigenous forest area — are supplying more than 80 percent of Kenya's industrial wood demands for both domestic consumption and export. The net effect has been to reduce the intensity of exploitation in the remaining 2 million ha of indigenous forest, the primary function of which remains that of catchment protection.

The second point is that, as part and parcel of this overall forestry development programme, the Kenya. Forestry Department, some 20 years ago, established 43 000 ha of nature (biotic) reserves. In the second forestry project financed in 1976, one condition of the loan was that these reserves would be extended by a further 7 000 ha, so that they would) become fully representative of Kenya's biological and botanical ecosystems. This was done.

A third point of general interest relates to Kenya's enlightened land-use policies in the area of forestry. Because of intense population growth and the fact that much of the forest is situated on soils of high agricultural potential, the indigenous forest areas have always been under pressure for agricultural settlement. In the fifties, a series of long-term comparative catchment area studies was carried out by EAFFRO4 to compare the impact on stream flow, soil erosion and downstream sedimentation of alternative land-use systems, including natural forest in an undisturbed state, plantation forestry, tea plantations, livestock and intensive food cropping. It was clearly established, given appropriate soil conservation measures, planting spacement and other husbandry techniques, that tea, for example, could provide an effective catchment cover without adversely affecting downstream flow and sedimentation. This long-term experiment was used as a basis for a deliberate decision by the Government of Kenya to excise some 10 000 ha of forest land in the southwest Mau Forest for subsequent tea production. Tea exports have now grown to be Kenya's second largest export earner after coffee, generating foreign exchange earnings which account for 25 percent of agricultural exports and 10 percent of total exports. Most of the tea industry which has enabled some 20 000 farmers to move from a subsistence to a cash-crop farming system is located on what used to be indigenous forest land.

TRANSMIGRATION PROJECT, INDONESIA

The Indonesia Transmigration Project, as the Colombian Caqueta Project, concerns the settling of small farmers on acid tropical forest latosols. In Indonesia, the emphasis is on arable crops, whereas in Colombia it is on livestock.

As part of a long-term transmigration programme, the Government of Indonesia requested World Bank assistance in 1973 for a project to help resettle incoming families from Java and Bali on four sites along the trans-Sumatra highway in the Province of Jambi and to upgrade the standards of living of existing families already settled at a site in the same area.

The Indonesia Transmigration scheme is one of the largest resettlement programmes in the world. Since 1905, successive governments have sponsored the migration of poor farmers from the overcrowded islands to relatively under-utilized neighbouring islands, particularly Sumatra. All told, government programmes have transferred nearly a million settlers, and an estimated 2 million Javanese residing in the Outer Islands are there as a direct result of government resettlement and associated population growth. Much of the earlier settlement provided workers for rubber estates in Sumatra.

In January 1974, FAO undertook a study to identify a possible transmigration project suitable for external assistance, and in 1976, based on the results of this study, the World Bank undertook a first-phase transmigration project intended to upgrade the living standards of 12 000 settled families and to establish a new community for 4 500 new settlers. New migrants were provided with five of land, of which 0.5 ha was already cleared and 1.0 ha already planted to immature rubber. A second phase project is now in progress, building on the experiences gained. A smaller farm size (3.5 ha) is being adopted.

The most controversial issue has been the question of the sustainability of the cropping pattern, taking into consideration the highly acid nature of the :forest latosols, deficient in nitrogen and phosphorus and possibly low in potassium. Earlier research showed that soil structure is favourable to plant root formation and that by adding regular fertilizer inputs some of the forest soils would become suitable for upland food crop production. To combat the high phosphate fixation, the initial phosphate application should be heavy. Nevertheless, no technical package involving a high degree of dependence on annual food crops has yet been proved over a long period of time.

The cropping pattern originally envisaged under the project allocated 3.5 ha of land per family, of which two ha were for food cropping and about 1.5 ha for tree crops (mainly rubber), the latter to be grown as a monoculture. Land clearing was to be carried out by a combination of mechanical and hand methods and 500 kg per ha of rock phosphate harrowed into the soil just prior to settlement. The main food crops to be grown were rice, maize and cassava and it was assumed that settlers would establish house gardens containing vegetables and tree crops such as coconuts, cloves, coffee and bananas and different fruit trees. Special provision was made in project design for ensuring that farmers would have adequate supplies of fertilizer, that there would be a framework for close coordination of the various government agencies :involved in providing extension support, and that seed and planting materials would be readily available for the farmers as and when needed. A staffing ratio of one agricultural extension worker per 500 families was planned, as well as a strong emphasis on training.

Despite these provisions, a recent review of project progress has highlighted the fact that incoming settlers are having difficulty in producing enough food crops to ensure subsistence and in securing the necessary inputs, such as fertilizer and improved seeds.

The key policy issue is whether there is any practical alternative to forest settlement in Indonesia in the light of increasing population pressure on the limited areas of good soil. The "alang-alang" grassland areas and the "Cerrado" region in Brazil, for example, could in theory provide a short-term alternative to continued forest settlement and allow more time needed to develop sustainable farming systems for the tropical forest latosols. However, in practice, a sustainable farming system for the "alang-alang" grasslands has not been developed. The scope for more intensive research in this area is a matter of high priority.

The question has sometimes been raised by environmental and other agencies as to why the World Bank supports such settlement projects in situations where there are significant ecological risks? Part of the answer is that spontaneous settlement as a result of population pressure is a fact of life in many tropical situations, has been going on for many years and in some cases is beyond government control.

By actively working toward improvement of existing farming systems, upgrading of extension services, assurance of a ready supply of agriculture inputs, and supporting more intensive agricultural research, the chances of preventing ecological degradation should be enhanced. The alternative -allowing spontaneous settlement to proceed unchecked - would leave farmers with inadequate inputs, and without extension services, roads, social services and marketing and other facilities.

As was noted earlier when dealing with the Malaysia Jengka project, it seems important to keep in perspective the fact that part of the remaining tropical forest ecosystems could be put to more productive and sustain able land use, for example, by conversing it to perennial agricultural tree crops and thereby providing thousands of small farmers with a viable alternative to shifting cultivation.

A SMALLHOLDER TREE-FARMING PROJECT, PHILIPPINES

The unique feature of this smallholder tree-farming project is that, with the exception of a project in Gujarat State, India, it is the only one -financed by the World Bank, to date, where small farmers are growing forest trees as a cash crop. This is a second-phase project and it has two :main components: smallholder tree-farming through a supervised credit scheme operated by the Development :Bank of the Philippines; and pine plantation development by the Bureau of Forest Development.

The smallholder tree-farming component is encouraging farmers on marginal agricultural lands throughout the country to take up tree farming (associated with food crop production) for establishment of firewood, charcoal, pulpwood and leaf-meal plantations. The project is innovative and experimental and is based on the :Bank's experience of an earlier US$2 million pilot project which provided funds for the development of pulpwood resources around the PICOP Pulp Mill. The first-phase pilot project was successful and has led to a quantified and readily perceivable improvement in the participating farmers' income and way of life.

Under the second-phase project, out of 28 000 ha of tree-farm development to be financed, 10 000 ha will be located in Mindanao, 5 000 in Visayas and 8 000 ha in the Ilocos region. of northern Luzon. Tree-farm size ranges from two to 15 ha. Fuelwood and charcoal plantations, which account for a high proportion of project farms, average about five ha.

In relation to the likely impact of the project on rural incomes, experience under the Philippines I Project is well documented and it would seem reasonable to anticipate sustainable net revenues of something between US$78 and US$100 per ha from tree farms of *Albizia* producing pulpwood, something in the order of US$ 140 per ha for tree farms producing fuelwood and charcoal, and US$300 per ha for farms producing leaf meal.

The financial rates of return to farmers are high and the project's economic rate of return is something in the order of 23 percent. The second project is experiencing difficulties related to land-tenure constraints, and the need for greater flexibility in this area is under review.

As for ensuring that forests are protected, the most interesting feature of the Philippines smallholder tree-farming experience is that it is mobilizing shifting cultivators in the reestablishment of forest cover in formerly degraded catchment areas. The profit incentive of tree farming is helping to encourage reforestation of eroded catchments.

Despite the obvious attraction of this formula, there are limitations to its wider application. One of the main problem areas in planning for expansion of the first-phase project proved to be the economic radius for haulage for pulpwood. Smallholder tree farmers situated outside a 100-km radius from the mill were excluded because of the transportation cost factor. Projects of this type are particularly suitable for establishment of plantations around a central processing plant (whether for the production of pulp, charcoal, power generation, alcohol, lumber or leaf meal) where there is a guaranteed market price for wood. But all of these different industries have upper limits of delivered wood cost beyond which it is not possible, profitably, to process the raw material. In other words, they are primarily suitable for concentrated resource development within the command area of a processing plant. For this reason, this approach could not be adopted as a "blanket" solution for all proposed forest areas in which shifting cultivation is a serious problem.

The scope for extension of the Philippines experience to other countries is, nevertheless, considerable and the World Bank is reviewing prospects for helping some of its other member countries to undertake similar schemes.

One of the most effective ways to slow down the rate of tropical forest destruction is to attack the root cause of the problem-rural poverty. If we continue to depend only on exhortations to logging companies, multinational corporations and developing-country governments to "stop tropical deforestation", it is my own personal view that we will be no more successful in arresting the pace of forestry destruction than was King Canute in trying to stop the advancing waves by using equally futile tactics. A deliberate shift in conservation strategy is needed to focus more on positive approaches to rural development and alleviation of rural poverty, if we are to preserve effectively what is left of the tropical forest ecosystems.

To save the tropical forests from further depletion the focus should be more on how to improve the incomes and quality of life of the 200 million who practice shifting cultivation throughout these forests.

AGRI-SILVICULTURE SYSTEMS IN UGANDA

Agri-silviculture is a production technique which combines the growing of agricultural crops with simultaneously raised and protected forest crops. This practice, called agri-forestry, has been in existence in various primitive forms since man learned to clear forests and cultivate land, and has different names in different parts of the world. In western and central Africa, the age-old habit of swidden agriculture, or "shifting cultivation," involving continued destruction of forest areas by cutting and burning and then raising the agricultural crops on the ashes of the destroyed forest, seems to be the beginning, however crude, of this practice. In Kenya they call it "shamba"; in parts of Europe and a large part of Latin America it is called "squatter planting". A similar practice involving forest villagers and tribesmen is known as the "taungya system" in Asia.

The origin of the taungya system can be traced back to 1862, when the colonial British employed Burma's taungya tribes for raising teak plantations along with their paddy (rice). In Indonesia, Malaysia and other tropical countries of southeast Asia, the practice is remarkably similar to that of shifting cultivation in Western Africa. In the local vernacular of Bangladesh the practice has been most appropriately called "jhooming," which means going round and round and never standing at one place; the farmers would change the site every year moving, year after year, throughout the available area. Various forms of shifting cultivation are still destroying forests, degrading land and causing erosion in many parts of the world on a vast scale. There are an estimated 3.6 thousand million hectares currently under shifting cultivation throughout the world, and it is believed that some 250 million persons live by it.

The present-day concept of agri-silviculture is in fact a way of reconciling conflicting interests of native cultivators and foresters and also is seen as a valuable and workable means of diverting the pernicious systems of shifting cultivation into constructive channels. This involves the establishment of an agricultural-cum-silvicultural cycle so as to use the labour of shifting cultivation to establish and maintain trees. It has come to be a valuable symbiosis and it is being accepted as a very popular system, wherever conditions are suitable, for promoting the economic and social development of the people living within or near the forest area, and simultaneously preventing the degradation of soil and vegetation.

The technique holds a significant position in tropical silviculture. As a form of land use, it aims at an integrated use of the available land resource to obtain a maximum amount of goods and services.

In various ways agri-silviculture seeks to create harmony between crop farming and tree farming. Decidedly, the system is not an easy one and at times the plans may not fully materialize. indeed, those patient, flexible and intelligent administrators who understand and sympathize with the forest farmers, their traditions and their human needs are among the most important elements of success of any system of agri-silviculture.

In Uganda, the system does not have a long history. It started as a form of tree farming when the hardwood species *Maesopsis eminii* (Musisi) was raised in combination with cocoa to provide partial shade to cocoa in its initial years of growth and to produce Musisi timber into the bargain. Later, a small-scale taungya system was applied in the Kifu forest to regenerate some indigenous trees. Taungya plantations were also developed for planting exotic pines in the Mwenge plantation, near Fort Portal, and at a few other sites with varying degrees of success and the system is being currently applied in the Sebei and Lira Districts where large-scale plantations of pine and cypress have been taken up under agri-silviculture. One such site is the Kachung Forest, the focal point of this study.

The Kachung Forest Reserve with an area of 35.4 km^2 is located 27 kilometres south of Lira town along the Lira-Soroti road in Uganda. The area has two peak rainfall periods and three to five months of dry season every year. The wet spells last from April to May, and from August to September. The relative humidity at midday is estimated to be 30-

65 % during the dry season and 60-95 % during the wet season. The temperatures are quite high, maximum 29.5-35°C′ during the dry season and 21-29.5°C′ during the wet season. The minimum temperature never falls below 15.5°C. The soil is mostly lateritic with reddish brown sandy clay loam at few places overlaying the laterite. Before its establishment as reserve this vast savanna area was being used as a communal grazing land. Subsequently in 1938, abortive efforts to plant *Chlorophora excelsa, Khaya grandifolia and Gmelina arborea* continued until 1950. The emphasis then shifted to trials of softwood plantation which showed their results around 1970. Soon after, fairly large-scale plantation work started with *Pinus caribaea,* and on a limited scale with *Pinus patula and Pinus oocarpa.*

There is no definite record indicating exactly when the agri-silviculture (taungya) practice was started in this plantation. According to verbal reports it is generally believed that it started around 1966/67 when the villagers living within the forest area, urged by some teachers from a neighbouring primary school, were encouraged to take up some land for cultivation by the Forest Department. They quite willingly took up the land, though on a quite e small scale. That state of affairs continued up to the end of the sixties when the Forest Department decided to expand the operation by developing a new strategy. The local village chiefs were approached to convince the local people around to extend cooperation to the Forest Department by taking up plots for taungya practice. At first the call met with a mild response from the local people, with just a handful of farmers joining the older lot. However, by 1970/71 there was a great inpour of people who took up plots. This was mainly due to the realization by the farmers that the area was agriculturally very productive. Since then the interest of the people has grown year after year.

Generally, the people use the land around the forest in two basic ways:

— Raising crops, mostly for food, such as millet (*Eleusine coracana*), pigeon peas (*Cajanus cajan*), cotton (*Gossypium* sp.), sorghum (*Sorghum vulgare*), simsim (*Sesamum indicum*), groundouts (*Arachis hypogaea*), several types of beans, cassava, sweet potatoes, various local vegetables and some sweet bananas.

— Keeping cattle, and some other domestic animals like goats, sheep and poultry. The cattle graze usually in communal grounds set aside for the purpose, away from agricultural areas. Because there is no risk of their trespassing into anybody else's cultivated garden, the cattle herds are let free every morning to graze on their own and are collected in the evening. The sheep and goals are not usually herded but are tied near the homestead except during the dry season when they are also left free. The use of grazing land is unplanned.

People work the land with a hoe -called locally the "lango hoe" because most of the inhabitants of the area belong to the Lango tribe. It consists of a flat heart-shaped blade fitted onto a rather long handle. Such hoes are exclusively used in the Lango area. Very few farmers possess ox-drawn ploughs.

There exists a customary land tenure system in the area, both for cultivation and grazing. The land usually belongs to a tribe, clan or family. In some cases it belongs to various individuals in the family, later inherited by their offspring, and this often leads to possible fragmentation.

Any land which has been lying unoccupied for ten years can be occupied by an individual, a family or a clan. The occupation must last for a long time and the land put into effective use during that period without anybody else claiming it, before ownership can be proclaimed. Despite the system of customary tenure, the lands under Uganda laws are known as public lands, vested in the land commission of the Government of Uganda., and are administered by the District Land Committees. Because of a rather light population pressure there seems to be no serious land problem in the area at present. However, there is a great hunger for fertile lands, which are shrinking.

The method of cultivation consists of clearance through repeated burning of all the existing vegetation, next digging, then sowing or planting, digging for weeding, harvesting and clearing the remains again. by setting fire.

The Langos have a communal system of land cultivation known as "Wang tic" in their language. The system carries many local nicknames in various parts of the Lango district. For example, around the forest reserve it is referred as "Alulu". This is quite an old form of cultivation. All persons, male or female of all age groups capable of carrying out cultivation work, belonging to a homestead or a group of homesteads, organize themselves into a "Wang tic". The leader of the group is chosen by the members. He is responsible for the organization of the cultivation, and fixes the order of cultivating the "gardens". The area covered depends upon the size of the "Wang tic". During the digging, each person is allocated an area to complete and the straggling members are usually helped. After every such cultivation the owner of the garden leas to provide local beer ("Kongo ceke" or "lacoi') or occasionally a meal.

After the demarcation of annual planting coupe, the initial clearing of land is carried out by the Forest Department toward the end of the dry season. This involves cutting down the entire existing savanna vegetation, except for a few valuable tree species. The cut trees are either sold as fire-wood or converted into charcoal and then sold. Usually fires are avoided as a precaution against any possibility of a serious outbreak of forest fire.

The planting operation is completed by the Forest Department before the farmers are allowed in. The forest species are planted at 3 × 3 metres spacing soon after a reliable rainfall, that is, when the wet season has really begun. The land allocation to farmers is usually done on a first come, first served basis. The maximum area allowed per tenant, usually a family, depends on the capability of the tenant which is judged by the size of the family and his social status. The bigger the family and the better its social status, the greater is considered to be the capacity of the tenant. While the allocation of plots is proceeding, tenants who wish to be neighbours can choose to have adjacent plots.

The tenants are not allowed to burn anything at any time in their plots. They can grow any annual crops but no perennials are allowed which are liable to cause suppression of the tree seedlings. The number of years for which a tenant can keep a plot are not stated at the outset. The maximum allowable period is dependent on the degree of crown closure of the forestry crop, the type of agri-crop to be raised and the degree of soil fertility. Ordinarily three to four years are required before the farmer can change the site. During the occupancy the farmer is required to look after the forest plantation.

The demand for land has been gradually but continually increasing. Most of the area taken up is along the Lira-Soroti road. A survey of tenants according to sex and age shows that a greater part-60 percent-of the tenants come from outside the forest area and are middle-aged women. The workers from within the forest area are, however, mostly young males.

To assess the effects of inter-cropping on the main tree crop, which is *Pinus caribaea* in Kachung, measurements and observations were taken in areas having almost the same age but have been under agri-silviculture for varying lengths of time. Because of the size of the available plots, 100 percent of the full area of each type was included in the study.

The agri-forestry practice has a beneficial effect on the condition of the forestry crop mainly through the elimination of the rank growth of weeds, which offer a very strong root and shoot competition to a sapling raised in an area where there is no agri-forestry or it has been practiced for only one year. Weeds in such cases are ineffectively removed and readily reappear. The most abundant weeds of the area are: *Imperata cylindrica* (sword grass). *Acacia hockii; Hyperrhenia rufa; Acacia mellifera; Grewia trichocarpa; Vangueria opiculata; Asparagus flagellaris; Combretum spp.*

Imperata cylindrica is a suppressant which tends to grow in pure forms excluding almost all other vegetation. It is very difficult to control during the first year of cultivation because of its sturdy rhizomes.

As for the effects of agri-silviculture on agricultural crops, information obtained from the farmers working in agri-forestry has been compared with answers from farmers working on other lands in adjoining areas.

All the agricultural crops tried so far under the system have proved promising and their yield is better than the outside land. The cultivation of agri-crops has to stop when the tree canopy above closes. There is a declining trend in production on both types of land due to the effect of soil leaching and exhaustion of fertility.

Taungya's Effect on the Local Ecosystem

Because the farmers clear the site of weeds through hard work, they would like the area to be available for cultivation for a longer period. This at times provokes feelings of hostility in the farmers toward the tree crop. Some of them would even physically uproot the tree while digging the adjoining land. Some instances of heaping weeds on top of saplings have been recorded. In other cases fires were initiated to kill weeds which had reappeared after the first year. But all such instances are exceptions rather than the rule and such tendencies are very much controllable.

There has so far 'been no serious attack of pests or diseases on either the trees or the agricultural crops growing together. In April 1977 there was however a grasshopper attack on millet. That appeared to be a region-wide attack, which was not restricted to the forest area, although some of' the farmers alleged that the attack was more severe in the forest area. There is also no evidence of attack by monkeys and browsers present in the forest reserve on the agricultural crops.

Though soil working had improved the aeration and water-holding capacity of the soil a small incidence of erosion was observed on slopes, especially on sites where the agricultural crop was still young. Nevertheless, local farmers and forest workers are of the view that replacing of the rank growth of weeds with agri-forestry crops had greatly improved the aesthetics of the area. The enhanced production of agri-crops and improved rate of growth of trees are an important socio-economic contribution to the area.

Generally, the people involved in the practice (the farmers and the Forest Department) are satisfied with the yield of crops from the land, the present land allocation system and the present way of working. The practice has evolved a sort of block farming. The Forest Department employees continually check their tree-crop though some of the farmers discourage them from passing over the cultivations. Because of the general economic condition of the farmers, mechanized farming seems to be a remote possibility unless some other agency takes the initiative.

Effect on Labour Costs

If agri-forestry had not been adopted, the Forest Department would be employing labour to carry out the entire tree-planting operations. For example, the planting coup of 1977 only, as calculated from past records, would have required 8120 additional man-days to establish the plantation of *Pinus caribaea* over 116.5 ha over the first three years. And in spite of so much investment the achievement would be only forest trees, which anyway are favourably affected under agri-silviculture. Agri-silviculture -or taungya— is a method seeking in various ways to create harmony between crop farming and tree farming. It is a compromise between two apparently conflicting demands. It is not easy and may :not always work as planned. But in the tropics, where the pressure of population is ever-increasing, such compromises are the only answer. Shifting cultivation and savanna land utilization have to be integrated to get timber, food, fuel and other allied products to ameliorate the economic condition of the developing tropical countries. It is believed that in the world today an estimated 3.6 thousand million hectares are under shifting cultivation and some 250 million people live by it. In Uganda alone about 10 percent of the total land area (its savanna) is ravaged by the fires caused by shifting cultivators or graziers each year. This calls for a change: change in. land utilization, change in attitudes and change in the ultimate fate of millions of people.

Ecological Factors

No doubt there is quite some amount of change caused by the planting of trees alone, without involving the taungya system. And the change is much greater if the tree rows are interplanted by agricultural crops. The rate and extent of change appear to depend upon the number of years of cultivation and the type of crop cultivated. In the Kachung area, after one year of cultivation there is a sudden vigorous re-growth of grasses, herbs and shrubs. *Imperata cylindrica* tends to come back in a big way, and after it has been weeded out again there is a quick invasion by lesser competitive species. That is why the areas that are cultivated only once may not ultimately be very different from those that are not cultivated at all. The vegetation changes become more and more conspicuous and permanent as the number of cultivations increase.

The constant and repeated cultivation has the effect of making the vegetation simpler, as the farmer makes incessant efforts to keep his crops weed free. And as the pines grow and close up, a new type of vegetational cover develops which is quite different from that in the neighbouring unplanted areas. The trees influence the environment through casting shade, keeping the temperature low, minimizing evaporation from the soil surface and affecting the microflora and microfauna of the soil. A new micro-environmental set-up comes into being. *Imperata cylindrica* cannot re-invade the area under pine shade as the grass is not shade tolerant.

As can be seen from the foregoing, the taungya practice has its beneficial effects on the growth of pines. The faster rate of growth is due to reduction in competition from grasses and other weeds. Most agricultural crops tend to be less competitive than the natural vegetation. Tropical soils are extremely susceptible to nutrient leaching, particularly when the ground is made bare of its vegetational cover by exposure either through cultivation or through other means. The soluble substances are continually washed down into deeper layers of soil. Some of these nutrients are drained down while the rest are taken back by the deep-rooted plant communities. Since it is only the pines that are a deep-rooted vegetation within the agri-silviculture area, they take back those nutrients. This in part may also account for the increased vigour in growth of the pines.

The agricultural crops, in their own way and at a given depth, also compete for nutrients with young pine seedlings. Millet, when looked after properly on a fertile soil and allowed to establish, forms a thick network of adventitious roots and a dense growth of parts above ground. It is therefore essential that the pines be given an initial advantage over millet, perhaps by delaying the sowing of millet or not sowing millet in the first year. Pines can comfortably cope with millet in the second year and onward. *Cajanus cajan* (pigeon peas) also need the same precautions. They grow to form a sort of shrub attaining a height of over 1.5 metres within three to four months and tend to overtop the pines in addition to making nutrient demands on the soil. However, an obvious result of the taungya practice is an increased yield per unit area of the agricultural crops during the first year. This is due to the still unleached fertility of the soil-the state of being uncultivated is a form of fallow and it is known that fallow is the best way to maintain tropical soils in the absence of fertilizers. The longer the fallow period, the better. The effect of allowing land to remain fallow results in striking increases in yield, regardless of fertilizer treatment. During the second year the yield of agricultural crop falls and from the third year onward it almost becomes constant. Obviously this decline in yield is due to a decline in the availability of nutrients. The yield can be raised if somehow the tillage can be taken deeper than the layer which is used during the first year. The cultivation can continue to about four years when the pines almost close up the canopy.

Effects on Socio-economic Aspects

Although communal farming has been in existence in the Lango area of Uganda for a long time, the system has got better organized thanks to the taungya practice. In traditional communal farming the shambas (gardens) are scattered, but the taungya system has brought them closer. Equity in site, working conditions and walking distance have gone a long way to re move unnecessary grudges and have helped the people to get better organized. This is reflected by the fact that at occasional parties - like at Christmas, Easter and other occurrences-usually organized within the forest reserve by workers, outsiders are welcome. Conversely the people living next to the forest invite the forest and taungya workers to their parties as a group. Block farming has decidedly contributed to the social development of the community. Millet and pigeon peas are the main staple foods of Uganda's north and they can be grown as agri-silvicultural crops.

The taungya system has a tangible effect on saving the operational cost of afforestation. Another benefit is the faster rate of growth of the trees resulting in more yield in shorter rotations. Not only the Forest Department gains directly, but it is the farmer who gains more from an increased yield of his agricultural crops. And this benefits is available on a sustained basis as every year new areas keep coming under agri-silviculture. That is why there has been a constant increase, every year, in the number of people engaged in the practice.

It should, however, be noted that the practice of agri-silviculture in Uganda has passed its infancy but is still young. Through careful planning and wise manipulation vast lands and human resources can be put in a progressive and productive combination. The communal farming habit of the people can be used very advantageously by educating them in agri-silviculture. Terms and facilities can be improved and the timing of planting the forestry and agricultural crops can be suitably adjusted. Research is required on suitable crop rotations and possibilities of introduction of perennial crops like cassava and dessert bananas. There are opportunities for helping the farmers by making tractor-hire services available, by arranging convenient sales of agricultural crops and by procuring basic agricultural tools and insecticides, at reasonable costs. Agri-silviculture needs to be carried out as a matter of policy and not only as a localized means to get free labour for forest areas.

AGRICULTURAL PLANTATIONS: FUTURE CONCERNS

Because of uncertainties in some of the traditional smallholding cocoa-producing countries, cocoa cultivation is becoming geographically more widely distributed and is increasingly being grown on a plantation basis, either as a monoculture or under coconuts. In Malaysia, the traditional tree fruits may follow a similar course and black pepper is also seen as a target for the specialist-producer.

This trend cuts right across the emphasis in current international forestry literature on the potential for "agri-forestry", that is, simultaneous inter-cropping of trees and food crops. It is important, therefore, to clarify the difference between these monoculture farming and forestry systems and the integrated food and tree inter-cropping farming systems practiced, for example, by small farmers in Java, in the Kerala region of India, in Sri Lanka (the Kandy Garden System), and so on. The very small farmer of the humid tropics with less than two hectares of land, typically grows a variety of food and cash crops around and near his house. In Java, the farmer is highly skilled and cultivates rice, cassava, maize, beans, groundnuts and vegetables in association with bananas, plantains, citrus, cloves, cinnamon, pepper, coffee, cocoa and a variety of tree fruits, all under a thin stand of coconuts.

The homestead tree lot, so typical of the humid tropics, reaches its highest expression in Sri Lanka, where the "tree gardens" round Kandy present a complex association of cassava, bananas, ginger, plantains and others under a mixed stand of tree fruits, coffee, cocoa, pepper, cinnamon, cloves, areca palm and coconuts. In West Africa, this sector is represented and vegetables are grown in association with a mixed stand of coffee, cocoa, tree fruits, kola and oil palm.

These small-farmer systems contrast sharply with the simplicity of mono-cropped agriculture and forestry plantations and by comparison would be more difficult to modify and improve. Improvement of smallholdings such as those in Java and Nigeria may only be possible by the provision of better planting material over an extended period of time, but a catalytic effect might be achieved by better roads and marketing facilities. These would stimulate the larger and more progressive farmers into modifying their cropping systems to take advantage of the better circumstances, but for the very small subsistence farmer, the element of risk could still be too high to permit change; under these circumstances, some degree of land consolidation and cooperative farming might be essential before improved cropping systems could be introduced.

As a broad conclusion, it seems that the "agri-forestry" farming systems used by small farmers in Java and elsewhere in the humid tropics are well proven and provide a diversified combination of subsistence, food and cash crops which reduce the risks of starvation and, at the same time, offer some small surplus cash income. However, it seems possible that where small farmers have room to manoeuvre and expand the scope of their cash-cropping operations, the trend is likely to be toward monoculture rather than away from it. In other words, in the humid tropics agri-forestry combinations may not always prove to be the most productive small farms.

The distinction between monoculture and inter-cropping (agri-forestry) farming systems deserves attention because recently there has been a tendency for foresters to jump on the agri-forestry bandwagon and promote indiscriminately agri-forestry systems in all areas of forestry development as a means of increasing the productivity of tropical forest lands. A more selective approach seems to be warranted with emphasis on those small farming systems or phases of development in plantation forestry where inter-cropping of food and tree crops can be of definite technical and economic benefit.

An associated issue is this: if we take a closer look at the potential for introducing agri-forestry in the humid tropics outside the well-established taungya plantation model and study the cropping pattern being used in such places as Java, Kerala, and Sri Lanka, we find that most of the trees which are being grown are fruit trees or horticultural crops which traditionally have fallen outside the foresters' province. Clearly, if foresters are to play a more active role in this area, we need to broaden our knowledge of the range of tree crops which can be used in forestry and to work in closer association with tropical agronomists who are familiar with such crops. We will also have to accept that this is an area in which the forester may often have to play a supporting role to the agronomist and agricultural economist rather than the converse. Investment in traditional forest tree crops will frequently, but not always, be a relatively low proportion of the cropped area and of farm investment costs.

A considerable part of the so-called "forest destruction" taking place in the developing world is, in fact, a logical shift in land use to more productive crop or farming systems. These are adequately supported with technical extension, agricultural inputs and other resources, land-settlement projects can provide a sustainable land-use alternative to retention of the land under virgin forest cover. There seems to be convincing evidence that many of the agriculture and rural development projects already initiated in tropical forest areas have resulted in a quantifiable increase in rural incomes; have enabled the small farmers involved to settle in more stable communities; and have eliminated their former dependence on shifting cultivation. In other words, settlement of small farmers and forest protection need not be mutually exclusive objectives.

In some of the projects undertaken in the past in which land-use and soil. capability surveys were carefully carried out in advance of settlement, and agricultural development channelled into the flatter lands, it has proved possible to exclude a large part of the remaining forest from agricultural settlement and this has remained unexploited. In other cases, inadequacy of forward planning, or too high a degree of dependence on non-enforceable forestry protection legislation, has failed to protect the forest. This means that project design must be flexible and take into account the needs and aspirations of incoming settlers or small farmers.

Because of the wide variation of tropical forest soils, climatic and physical conditions, it is impossible to generalize about appropriate farming systems for tropical forest areas, but what seems to emerge from this analysis is that perennial agricultural tree crops such as oil palm, coffee, rubber, cocoa, tea and coconut can be an ecologically sound alternative to natural forest management and, secondly, that whatever agricultural crop or livestock or forest plantation crop combinations are envisaged, the capacity of governments to ensure adequate support services, inputs, such as fertilizer and seeds, agricultural research services, feeder roads, social infrastructure and marketing outlets constitute a decisive factor in determining whether a particular farming system is sustainable. Even in the most intractable soils, such as those of the Amazon, evidence suggests that, given appropriate attention to soil conservation measures and crop husbandry techniques, it may prove possible to sustain arable, and even livestock farming, systems on at least the better endowed of these low-fertility soils where, in the past, all such attempts have failed.

Because market constraints for plantation grown agricultural and forest tree crops will probably limit their development to something less than 10 percent of the remaining tropical forest ecosystems between now and the turn of the century, high priority should be given to agricultural research and to pilot-scale development programmes aimed at improving the present state of knowledge of sustainable food cropping systems as an alternative to shifting agriculture and, in the interim, directing settlement to better soils.

In watershed areas, an integrated rural or "area" development approach which offers the small farmer an alternative to his ecologically destructive way of life can help to preserve the remaining forest and, thereby, reduce the risk of soil erosion and downstream flooding. Investment in such infrastructural "inputs" as a supply of seeds and fertilizers, torrent-control structures, soil-conservation measures, provision of credit, training of extension staff, feeder roads, marketing services, schools, shops, hospitals and other social services is the quickest and most certain way to ensure that farmers abandon shifting cultivation and adopt a more sustainable farming system. Some agricultural and live-stock farming systems with appropriate husbandry practices can also provide effective catchment protection. In other words, forestry is not the only solution.

It is clear that compensatory forest plantations have an important role to play in ensuring protection of part of the remaining tropical forest ecosystem because they can provide an alternative source of timber and take the pressure off further exploitation of indigenous resources. Reforestation programmes in the developing countries are currently proceeding at less than 20 percent of the rate needed to ensure domestic self-sufficiency by the year 2000 and a massive increase in the annual rate of establishment of fast-growing species is called for before smallholder tree-farming can play a significant role in situations where the forest lands or catchment areas to be protected are situated within an economic haulage distance of a processing plant or cash market.

Concerning the problem of the preservation of biotic reserves, the environmental agencies of the world has e done a masterly job in alerting international awareness on this matter. It is becoming widely accepted that the arguments in favour of preservation of forest-dwelling hunter/gatherers, wildlife, botanical genetic resources and potential future drug and medicinal plants are irrefutable and the governments of some developing countries (e.g., Malaysia and Kenya) have expressed their willingness to increase efforts to protect such resources and have created environmental agencies. But the only certain way we have of ensuring that these designated biotic preserves will be protected in practice is by increasing support for rural and agricultural development programmes in adjacent areas.

Foresters might regard some of the projects described here as agriculture rather than forestry projects, and it is precisely at that point that problems can arise. Foresters in the past, have tended to be highly parochial about defining what constitutes a forestry project and to assume that their responsibility starts and ends with the cultivation of forest trees in forest reserves.

In fact, forestry investments are likely to comprise a relatively small proportion of many of the rural and agricultural development projects which will be needed to bring the current process of tropical deforestation under effective control. In the area of catchment and agri-forestry, in particular, foresters will have to be prepared to work more closely with agricultural settlement and other agencies, playing a complementary and supporting, rather than a predominant, role in the development process.

Concerning the World Bank's role in forestry projects, we are conscious of the fact that the Bank's efforts in this area can only be marginal and that the main impetus must come from within the developing countries themselves. Following publication of a Bank Forestry Sector Policy Paper in 1978, the Bank made a major shift in emphasis of forestry lending toward environmental and rural forestry, and set as a goal a five-fold increase in the :level of forestry lending to achieve a target of US$500 million within the five-year period 1979-83.

The response of the developing countries' forestry services has been encouraging. Since 1978 we have made loans to forestry projects in about 35 different countries, and more than 60 percent of these have been for programmes aimed at environmental protection and provision of fuelwood, fodder, building poles and other forest products needed both for basic subsistence and development. The US$500 million lending target has been achieved somewhat ahead of schedule.

REFERENCES

Curtis, A and Race D, *1998 Links between Farm Forestry Growers and the Wood processing Industry. Lessons from the Green Triangle, Tasmania and Western Australia,* RIRDC publications no 98 /41

David, Soniia 1995. *What do farmers think? Farmer evaluations of hedgerow intercropping under semi-arid conditions.* Agroforestry Systems 32: 15-28.

Dwivedi, A.P. and Sharma, K.K. (1989). Agro forestry: its potential. Paper in Seminar on Eucalyptus, FRI, Dehradun.

Eastin, I. 1996. *The role of marketing in promoting a sustainable trade in forest products in tropical regions.* Paper presented at the FAO/UPLB Workshop on Forest Products Marketing, Los Banos, Laguna

Falconer, J. & J.E.M. Arnold. 1991. Household food security and forestry. *Community Forestry Note No. 1*, FAO, Rome.

Mayadill, H.T., 1991, Agroforestry for Tropical Rainforests. Institute for Tropical Rain forest and Ecology. Agroforestry systems. 13(3):259-267 p.

Nair, P.K.R. 1979. "Agro forestry Research: A retrospective and prospective appraisal". Proc. Int. Conf. International Cooperation in Agro forestry. ICRAF Nairobi, pp. 275-296.

Olofson, H. 1983, *Indigenous Agroforestry Systems. Philippine Quarterly of Culture and Society,* II: 149-174.

Schirmer, A. 1983, *The role of agroforestry in the Pacific. Suva, Fiji, Univ. South Pacific.*

Shah, P, 1988, "Potential of Agroforestry as land use technology for rural poor". *Advances in Forest Research in India,* 1:27-100 p.

Tewari, D.N. 1995. *Agro forestry for increased, productivity, sustainability and poverty alleviation.* International Book Distributors, Dehradun.

8

Agroforestry in Semi-arid Tropics

In recent years, the attention focused on challenges facing the tropical forests has intensified dramatically. However, nearly the totality of this attention has been directed toward the forests of the humid tropics. Even in technical documents, it is not uncommon to find statistics on the extent of tropical forests and the rates of loss or alteration followed by narrative that deals almost entirely with tropical rain forests. Much less—in fact, almost negligible—attention has been focused on the challenges of forest resource management in arid and semi-arid areas.

Yet the problem of dryland degradation is one of the most serious resource management problems facing the world today. Occupying about one-third of the world's land surface area, and supporting a population of more than 850 million—almost 20 percent of the world total—the drylands are rapidly being degraded through population growth, overgrazing, cropping of marginal lands, inappropriate irrigation and, not least, deforestation. In 1983 the United Nations Environment Programme assessed the implementation of the 1977 UN Plan of Action to Combat Desertification and found that at least 75 percent of the productive land in arid and semi-arid areas was already at least moderately decertified (having lost more than 25 percent of its original productivity). The problem is particularly serious in the developing regions, where degradation of rangelands and areas adjacent to the borders of semi-arid rain-fed farming areas continues to accelerate. Although croplands comprise only about 15 percent of the total dryland area, they are home to more than 85 percent of the dryland rural population, and the pressure is growing. Deforestation is intimately associated with dryland degradation as the destruction of woody vegetation for fuel and other uses far outstrips the rate of replanting and regrowth.

In the late 1970s end early 1980s, forestry in arid zones tended to concentrate on the establishment of large-scale plantations of exotic species, aimed exclusively at timber or fuelwood production. To do this, large areas of the "useless brush" on which local people depended for fuel, food and fodder have been cleared. In the development of agricultural schemes or irrigation projects, natural tree formations have often been eliminated, leading to increased erosion and subsequent losses in productivity and shortened life spans of reservoirs. However, although the overall trend is still unsatisfactory, there is growing evidence of a change in thinking on the part of foresters and land-use managers concerned with arid zones.

CLIMATE OF SEMI-ARID TROPICS

The semi-arid tropics (SAT) cover an area of about 20 million km^2. Kampen and Burford estimated that 700 million people live in this zone, nearly half of them in India. This population undoubtedly has increased, perhaps by as much as 50 percent, in the intervening 11 years.

The SAT cover most of West, East and the southern part of central Africa; most of India, northeastern Myanmar, northeastern Thailand and northern Australia; most of Mexico; and large parts of eastern and central South America. The SAT environment is characterised by high atmospheric water demand; a high mean annual temperature (>18°C); and a low, variable annual rainfall (400 to 1900 mm). The climate of most of the SAT is monsoonal, with over 90 percent of the rainfall occurring in the period of April-October in the Northern Hemisphere and October-April in the Southern Hemisphere. This article focuses on agroforestry in the dry semi-arid tropics, i.e. those areas in which rainfall exceeds potential evapotranspiration for less than 4.5 months of the year.

AGROFORESTRY SYSTEMS

Parkland Systems

Although no definitive estimate exists, it is probable that the parkland system, characterised by mature trees dispersed in cropped fields, is the largest single agricultural land use in sub-Saharan Africa. Across the entire Sudano-Sahelian zone of West Africa, one finds crops planted under varying densities of mature trees. It is quite possible that some of these older parklands are remnants of elaborate pre-colonial agricultural systems, with cropping intensities and input levels not found in the region today.

The ability of these parklands, or two tiered systems, to enhance and stabilise crop production has been much studied over the past 20 years in West Africa and to a lesser extent in arid parts of India. The systems that have received the most attention are the *Prosopis cineraria*/millet mixtures of eastern Rajasthan, India, and the *Faidherbia albida*/ grain systems predominating throughout the Sahelian zone and in some parts of East Africa.

Many authors have shown an enhanced effect of these species, particularly *F. albida*, on grain crops growing underneath. Estimates of increased yields range as high as 100 percent compared with crops grown away from the trees. While this effect is generally attributed to improved soil conditions under the tree because of litterfall, recent evidence suggests that much of this fertility may in fact precede the tree. There is also new evidence that emerging crops benefit equally from lower soil and leaf temperatures under the light shade of the tree, which is defoliated at the start of the rainy season.

Figure 1. An Agroforest

Although there are no comparative yield data for *P. cineraria*/cereal associations from India, it is believed that crop yields are also higher when grown under this system. Limited evidence suggests that the tree has an unusual rooting pattern that does not seem to compete with crop growth. Muthana *et al.* excavated a 20-year-old *P. cineraria* tree to a rooting depth of 7.5 m. The horizontal extension of lateral roots of this specimen was less than I m from the trunk. The local practice of lopping the trees annually for the highly valued fodder soon after crop harvest ensures that the canopy never gets large enough to shade subsequent crops. Despite the fact that such traditional systems are beneficial, it is rather difficult to improve on them or extend them into other areas.

There are two reasons for this: the first is that other tree species create high rates of competition for crops; the second is that it has often proven difficult to establish these systems in areas where farmers are not accustomed to them. There are not believed to be definitive reports of species in SAT other than *P. cineraria* and *F. albida* that exhibit the ability to increase crop yields under their canopies. In fact, Prajapati et al. observed suppression of yield by *Prosopis juliflora* growing near a cropped field. Roots of the trees extended 30 m into the field, reducing sorghum yields by 80 percent over this range.

Similar crop yield reductions can be readily observed under most other tree species found in cropped fields of the SAT. However, in some cases, the *Parkia biglobosa*/cereal mixtures of the West African Sudanian zone, for example, crop losses are more than recovered by products produced by the trees. Canopy management may be one way to decrease this competition. Mann and Saxena stated that a sixfold increase in mung bean was obtained when it was grown under 12-year-old *Acacia tortilis* spaced at 4x4 m when the trees were pruned. Singh et al. showed the importance of intensively lopping competitive trees in tiered systems.

Although plots under pollarded *Leucaena leucocephala* spaced at 2x6 m had just half the sorghum grain yield of sole sorghum plots, these yields were ten times those in lightly lopped treatments. Therefore, lopping could be used to expand the range of species suitable for parkland systems, and the prunings could be used for organic fertiliser, mulch, fodder, fuel and other purposes.

Projects that have attempted to establish parklands by extensively planting trees on regular grids have not been successful. This is because of variable and/or slow tree growth and often a lack of meaningful farmer participation and interest.

Some success has been met with natural regeneration schemes, where farmers are provided incentives to protect young trees that emerge in their fields. However, this reduces control over tree species, seedling location and spacing. Given the high microsite variability in many SAT soils, an alternative strategy may be to identify "islands of fertility" in farmer's fields and plant locally favoured species on them.

The subsequent rapid growth and high survival rate would be attractive to farmers and would lessen the amount of time needed to protect small young trees. Another possible option would be to introduce more valuable cash crops in the favourable microenvironment found under the shade of existing mature *F. albida* or *P. cineraria* trees.

Wind-breaks

Tree rows as wind-breaks have long been used in semi-arid temperate regions of North America, Europe and Asia for protecting crops and soil against wind damage and wind erosion. More recently, their effectiveness in increasing crop production has been demonstrated in the drier parts of the SAT, particularly sub-Saharan Africa.

The few critical studies that have been done in the SAT confirm findings of increased crop yield behind wind-breaks in semi-arid temperate regions. In northern Nigeria, Ujah and Adeoye are 14 percent average increase in millet yield behind a *Eucalyptus camaldulensis* windbreak. In the arid zone of India, Sur reported a 21 percent average increase in yield of protected cowpea over six years. El-Kankany reported a yield increase of 36 percent for cotton, 38 percent for wheat, 47 percent for maize and 10 percent for rice in Egypt, where over 100000 ha of croplands are protected by windbreaks. In the Majjia Valley of the central Niger, double-rowed *Azadirachta indica* (neem) wind-breaks spaced 100 m apart showed a 20 percent increase in crop yield in two separate studies.

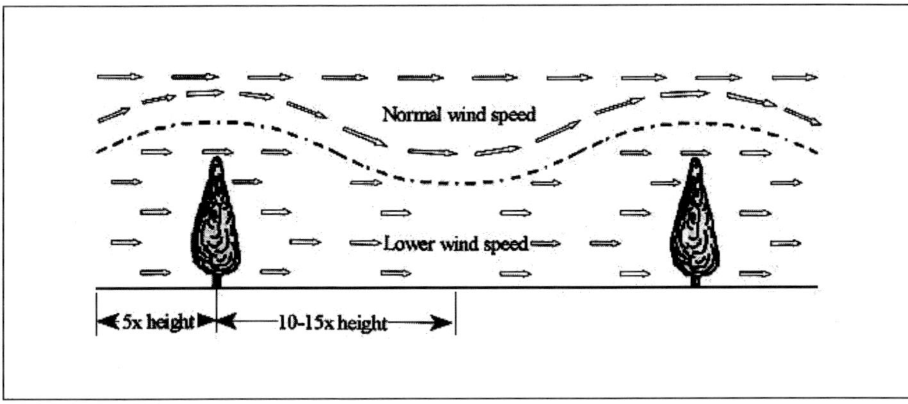

Figure 2. How windbreak works

The great interest generated by these data and experiences led to an increased emphasis on wind-breaks in West African development assistance projects during the late 1980s. These invariably met with mixed results. A major problem was uneven tree growth in wind-break lines caused by soil variability, which creates unevenness in wind-break heights and reduces their effectiveness. Too much faith was put in results from trials located on research stations or from sites of very successful wind-break projects. An example of this was the Majjia Valley project, where the trees benefited from a very shallow water-table which is unrepresentative of most other agricultural lands of the Niger.

Finally, wind-breaks in many of the projects were designed to run in straight lines perpendicular to damaging winds. From a technical point of view this is ideal, but in practice it can lead to inequable losses among small farmers of lands occupied by wind-breaks. Competition between the trees in a windbreak and the crops they are supposed to protect can also be a problem. In a study at the University of Agricultural Science (Bijapur, Karnataka, India), safflower was planted on either side of a line of six tree species. In order of decreasing competition, the species were *Acacia nilotica, A. catechu, Eucalyptus camaldulensis, Dalgergia sissoo, Leucaena leucocephala, Causarina equistifolia.*

Other than species selection, there are few management options for reducing underground competition. Kort advocates root pruning in temperate regions to a depth of 60 cm close to the wind-break. This is hardly possible in less developed nations, where draught power, even animal draught power, is lacking. As with two-tiered systems, canopy management may be effective in reducing competition for solar energy, provided it does not interfere with the aerodynamic efficiency of the windbreak. This was accomplished in the Majjia Valley by pollarding one of the two windbreak lines at a height of 2.5 m. The wood harvested from this operation has provided income for the cooperative which manages the wind-breaks. The successful establishment, protection and management of the Majjia Valley plantings have required nearly 15 years of sustained effort and funding from the Government of the Niger and donor agencies.

Similar long-term commitments from donors and governments are rare among development projects in general, let alone wind-break projects. Therefore, it is imperative to select sites wisely, i.e. those that will give a reasonable return on investment for wind-break establishment. Further research into the ecophysiological processes, favourable and unfavourable, that govern wind-break/crop interaction in the SAT is important. Imaginative ways to interact with small farmers in anticipating costs and benefits of such projects and in actually distributing them are just as necessary in wind-break extension as in that of other agricultural technologies. Another factor that bears consideration is the effect of wind-breaks on the air temperature on the down-wind side of the trees. A number of studies indicate that increases in air temperature in the lee of shelter-belts of semi-arid and arid areas may be as much as 5 to 10°C and may shrivel or burn crops.

Alley Cropping

Based on positive results from the humid tropics, particularly at the International Institute of Tropical Agriculture (IITA) in Ibadan, Nigeria, alley cropping has generated much interest among donors. This is particularly so in West Africa where a collaborative network has been established by the IITA and the International Council for Research in Agroforestry and under various national programmes. It remains to be seen whether alley cropping, which has been so successful on-station, will be widely adapted by farmers in West Africa, particularly in and to the north of the Sudanian zone where a reduced incidence of trypanosomiasis now allows extensive animal husbandry.

The fodder needs of these animals will compete with the use of hedgerow loppings for green manure, a key component of the alley cropping technology. Furthermore, it is now clear that alley cropping has definite drawbacks in areas where water is a limiting factor. Narrow hedgerows, spaced at distances recommended for the humid and sub humid tropics, are particularly competitive with crops in the SAT. Trials in India have shown that grain yield of millet grown in 3.6 m wide *Leucaena* alleys systematically decreased with alley age from a yield reduction of about 17 percent the first year to over 80 percent by the fifth year. This is probably because of underground competition for water.

Despite these limitations, the concept of carefully placed and closely pruned rows of trees in farmers' fields may have a role to play in SAT farming systems. Such vegetative barriers could be located on field boundaries, providing feed for livestock and delineating cropped fields. Similarly, if situated at suitable contour intervals, they may serve to protect highly erosive soils. Finally, as low wind-breaks, they could reduce wind erosion and actually trap saltating soil particles on sandy soils.

The future role of alley cropping in the SAT will depend on the development of management methods to control the sources of competition between hedgerows and crops, and a greater selection of woody or perennial grass species to use in the hedgerows. A better chance of success is likely if alley cropping is only considered for areas where the assured annual rainfall exceeds 750 mm on soils with good water holding capacity.

Silvipastoral Systems

There are many animal husbandry systems in the SAT, but two extremes may be recognised. The first is where animals are a permanent fixture in the farming system, and are primarily stall-fed with supplemental pasturing. The other extreme is the nomadic and transhumant systems of the Sahel, where animals are moved from one site to another according to changing pasture conditions throughout the year, or herded to distant pastures during the rainy season to protect crops. In both cases, livestock play an important role in sustaining field fertility through manure production.

In India, farmyard manure is normally collected and spread on the fields. In the Sahel, herders are paid to bed animals on arable fields. In both systems, livestock pressure must be balanced with rangeland vegetative productivity if it is to be maintained. However, most grazing lands in the SAT of Africa and India are public lands and are therefore freely accessible, a situation that may be leading to their overuse and irreversible decline.

Under communal ownership, the incentive is to expand one's herd, without concern for the effect it has on the land, in order to reap maximum short-term personal gain. At the same time, the long term impact of such individual misuse is shared by all users, and therefore the individual's long-term loss is buffered. In this situation, there is no personal incentive to make improvements on the pasture.

In India, this decline of communal pastures is exacerbated by the growing tendency toward private ownership of grazing land, forcing poorer livestock owners to share less and less land. In the face of this emerging regional crisis, it is clear that meaningful changes in land-use policy and policy enforcement is an important first step. Before this is done, agroforestry interventions, or any other production-side intervention, will have little effect on a broad scale. However, they can be useful on smaller scales: excellent examples from semiarid Australia, India and elsewhere demonstrate the usefulness of fitting the potentially excellent feeding value of SAT trees and shrubs into pastoral and agropastoral systems.

It is possible to think of all sorts of grass/tree/crop designs to produce more fodder at the farm level, a task far beyond the scope of this subject. However, some guidelines can be proposed.

First, fodder trees and other suitable perennial vegetation should be considered as components of almost every agroforestry system for the SAT. Examples would be the use of fodder trees in wind-breaks, or the management of parkland systems to produce a fodder component through pod production or lopping.

Second, more effort must be made by scientists to incorporate studies on increasing fodder production into farming systems research. Finally, the overall direction of research and development work at the farm level in this area should be the designing of efficient cut-and-carry systems.

Miscellaneous Systems

Where irrigation is available, or in river valleys where water-tables are close to the soil surface, many traditional agroforestry systems closely patterned after those in the humid tropics can be observed. Such sites offer rare opportunities to maximise the output of tree products on a year-round basis and to take advantage of their many uses as vine supports, living fences, shelter and the like. In non-irrigated areas, tree planting for specific farm management objectives is little studied but of significant potential.

Vegetative fences, either live or in the form of thorny prunings, are widely used in the SAT. Similarly, boundaries between fields and farms could be made productive by tree or shrub planting for browse and other uses. Vegetative livestock driveways are sometimes established in and around villages in the Sahel to control animal movement and define village boundaries.

Trees can play an important part in river and stream bank and waterway stabilisation efforts, vegetative contour strips for soil conservation and simply for shade. Tree plantings can fill production "gaps" in farming systems by making use of farm areas that are not usually cropped—for example, bunds, field corners and rocky and unproductive field sections.

MANAGEMENT OF ARID-ZONE FORESTRY

Both theory and practice show that in areas with a long dry season, even those which are nearly deserts, natural woody formations and perennial plants make the best use of the soil and climate. In addition to their capacity for self-protection and biological improvement of the environment, they also afford protection for more modest plants growing under their shelter. The supreme example of a plant which is both woody and perennial is the tree. Trees, so often relegated to the most mediocre soils or cultivated only as a subsidiary activity, have an essential. if not preponderant, function to perform in semi-arid and arid zones. They should provide the framework and even, in some cases, the very basis of the rural economy.

The arid lands in the warmer climates of the world are generally characterized by low and erratic precipitation, diurnal fluctuations in air and soil temperatures and low humidity, although dew may contribute substantially to available moisture in certain seasons. Winds may be strong, days very bright and sunny and extended periods of drought are unpredictable, but inevitable. This means that such lands have a low threshold of susceptibility to misuse. When basic thresholds of resource tolerance to human and animal pressure are exceeded, He result is either deterioration or destruction of the physical environment with proportionally adverse effects on the rural population.

The primary reason for the massive human tragedies that are so common in arid and semi-arid regions is essentially land misuse under the marginal and fluctuating climatic conditions particular to these areas. Successful rehabilitation and improvement of the land require elimination of abuses by man. The starting point for improvement of agricultural productivity in arid and semiarid zones should be integrated planning in which the real nature of the land and the real nature of the people who are a part of that land are taken into consideration.

One of the most common and unrecognized misuses of arid and semiarid ecosystems involves the cultivation of annual plants without conservation measures.

Annual plants degrade soil fertility and are marginal for two reasons. First, the position of their root systems in the upper layers of the soil exposes them to the risk of desiccation, and second, because of the unreliability of the climate it is impossible to respect their specific requirements which are sometimes very strict with regard to installation, growth and maturation. Moreover, owing to the unreliability of the climate and the very nature of annual plants, the soil remains without protection and becomes exhausted during both the growth of the plants and the dry season. When planted with annual crops, arid and semi-arid ecosystems that are inherently marginal pass quickly from a state of fragility to one of total exhaustion. In other words, they are not being used in the way they were meant to be used. Sedentary and stable forms of agriculture are simply not feasible with the introduction of exotic annuals.

With perennial plants the situation is immediately different. Their growth does not depend on rain starting within two weeks of a certain date. Moreover, they cover the soil better and last longer, a trait which represents a step forward in the stabilization of the communities and of the environment. Whether natural or introduced, perennial woody plants are more profitable than the annuals, since they can provide a whole range of products: fodder, wood, minor products, even food for people and animals. First in importance among the perennials are trees of all kinds-fruit trees, semi-forest and forest trees—which are able, owing to their diversity, to adapt themselves to varying requirements. If the conditions for their installation and management are met, these lefty, bushy plants can simultaneously produce fodder, food and energy, provide shelter, break the force of the wind and possess all the advantages of self-protection. Their roots help to deepen and improve the soil; the shade they provide facilitates ecosystem metabolism; and they afford to plants less endowed by nature some of the benefits of the protection which they themselves enjoy.

With these functions in view and considering how best to combine the various elements necessary for the development of a stabilized agricultural system in arid zones, trees come immediately to mind as being essential for guaranteeing this stability and ensuring the continuity of agricultural activity. Such activity can be divided into three groups, based on the different ecological conditions prevailing in arid zones:

- Extensive stock raising, of a nomadic type, based on the use by wildlife and domestic animals of vast stretches of desert ranges.
- Grazing and forestry activities in arid areas, concentrated on natural grazing land.
- Agri-silvi-pastoral activity in semiarid zones, combining arboriculture, crop growing, wood and animal production.

Farming in arid zones should be confined to areas where there are sufficient groundwater resources (springs, wells) or where it has been possible to concentrate surface waters by runoff. This also applies to trees, whether forest or fruit trees, because for trees the limiting factor is lack of water: either they produce a lot of wood and also consume a lot of water, or they consume little water but then also produce little wood. This either slows their growth or reduces them in number to a very few per hectare. Since the climax of arid zones is a shrub steppe, the possibilities for introducing trees of necessity are very limited. Rather than planting trees which grow upward, it would be better under these bioclimates to cover the soil widely with low vegetation to stop the sand being blown away, to avoid the formation of zones of accumulation through wind erosion, and to make the ground more rugged. A plant cover with the same characteristics as the original steppe should be reconstituted.

In semi-arid zones there are greater possibilities for crop growing and trees have an important role to play in stabilizing agriculture. For this they must either be permanently present or regenerate without interruption in the same area. In the first case, we would have permanent trees or shrubs either in rows or scattered about. Their role can be summarized as follows:

How have desert peoples, whom outsiders might consider c; primitives, come to terms with their environments? The answer is in the cohesive social systems of these groups which have enabled them to create and organize farming appropriate to their land and climate.

- Trees in rows protect crops against wind erosion and desiccation.
- Trees intermingled with crops protect the crops and reconstitute and enrich the soil.
- Thus improved, the cultivated zones lose their marginal character, rotation cropping becomes possible and a supply of fodder is ensured.
- The ideal combination of semi-forest trees bearing edible fruit, fodder trees and forest trees ensures not only a diversified and stable production, but also agricultural and biological balance.

In the second case, where trees are regenerated, they are grown during the fallow period as part of rotation cropping, in order to regenerate the soil and provide cattle with supplementary fodder.

Land-use Management

The outline of an overall land-use management strategy thus begins to take shape. Shelterbelts of forest species would be reserved for the great plains, where division of the land into plots and the presence of active farming prevent the close juxtaposition of trees. The hillocky peneplains, hills and slopes would be protected by scattered vegetation and would, over most of their area, be converted into a network of plots to be cultivated under a system of rotation cropping based on the use of woody vegetation.

Land-use management plans and forest policies in arid and semi-arid countries to date have been designed for and implemented in well-defined, individual areas where the tendency of the forester was to protect and conserve the forest for its own sake rather than adapt it to the requirements of the rural people. This attitude may have been justified in the past but cannot be accepted today in view of the accelerated degradation of the soils, the marginalization of agriculture and the growth in population. The basic prerequisite for an improvement in rural life is stabilizing agriculture, and trees should not be confined to areas marked off as marginal. Forestry should be considered not as a residual form of land use but as an indispensable component in an overall land management plan. Its presence guarantees conservation of the environment, improved renewal of the natural resources and existence of the conditions necessary to enable rural people to live a satisfactory life.

This being agreed, the stabilization of agriculture then becomes the forester's main contribution, rather than trees, shrubs, erosion control or the production of fuelwood. The whole art of the forester, which is essentially an art of management, really consists of incorporating silviculture into management of the whole land area and encouraging the adaptations necessary to ensure a greater contribution to the stabilization of land use.

In most arid and semi-arid countries the problem of land. management and the incorporation of trees is affected by the existence of traditional social and agrarian. structures. The fragility of the environment and the variability and shortage of rainfall have led to judiciously adapted structures geared toward protection of the environment as providing the only guarantee of existence. Serere agriculture offers a good example of the degree of adaptation which has been attained in arid bioclimates. The Sereres are peasants living in western Senegal, a region with poor soil and low, variable rainfall. They have succeeded in maintaining the fertility of the soil by planting useful trees in the cultivated fields, making regular use of the bush fallow system and adding manure. They have managed to feed themselves under difficult conditions while maintaining a stable environment even with a dense population.

Other examples can also be cited, such as the Dogoan and Hausaland regions in northern Nigeria, where population has reached a high density without degrading the environment, thanks to the fallow system, crop growing under the shelter of trees, and the use of manure.

How have these peoples, to whom outsiders might-mistakenly-apply the term "primitive," been able to come to terms with their natural environments, even conserving these environments? The answer may be found in the cohesive social systems of these groups which have enabled them to create and organize appropriate ways of farming.

Whatever strategy is adopted for the management of arid areas, enduring social structures such as these should not be neglected. Enhancement of these structures would bring nothing but benefits. Simple judicial or legislative changes would suffice to ensure that these structural groups, which are in fact management and action units, be officially constituted and recognized as the key element in land management. In addition, simple improvements of traditional techniques, which can be understood as mastered by the population, are very important. These should be adapted to local ecological conditions and should require only a minimum of imported products or experts.

The role of technical assistance would be to provide continuing advice, information, training and follow-up in the field on all operations involved in the establishment, management and utilization of the plant infrastructure and soil improvement. The essential requirements are to understand the needs of the people and ensure their active participation in forestry activities. This presupposes a real dialogue with the rural communities in order to make them truly aware of the contributions of trees and forestry activities and of the necessity to integrate them into plans for the management of their land. This dialogue, this information, this training must be based on a thorough knowledge of social and economic structures and of the requirements of the rural communities.

The technical abilities required for this type of assistance are certainly much broader than those traditionally needed to manage a protected area, for instance. Henceforth, technical experts will have to find ways of incorporating trees into land-management plans and other forestry activities with the technical support which will enable them to face up to growing, and changing, requirements. This will require a multipurpose orientation which will take into account the objectives of agricultural stabilization and wood production, the necessity to conserve the environment and the desirability of full integration into rural development.

It is essential that there be close cooperation among the various services to which the technical personnel are attached and the agencies working in planning, assistance and distribution of facilities. If necessary, increased responsibilities should be delegated within the forestry administration itself. This would lead to a progressive reorganization of the forest services and the recruitment of non-foresters (specialists in animal production, agriculture, veterinary science and sociology who could collaborate with foresters and form a truly versatile forest service.

Planting trees to stabilize agriculture is more costly the more arid the climate and the more degraded the soil. What is really involved is reconquest of the soil based on the establishment of a ligneous infrastructure. Since this represents an enormous expenditure, it can be easily understood why there should be hesitation about undertaking it. It will, however, have to be undertaken sooner or later. It is not sufficient merely to mobilize the local workforce for multipurpose plantation work after education and active propaganda; they must also be given all. the help necessary to obtain for themselves the means necessary for this new activity. In other words, it is necessary to provide assistance based on an obligation on the part of the state and commensurate with real requirements. This assistance may., according to the circumstances, be limited to education, organization and demonstrations, or it may take the form of material assistance: donations, subsidies and loans of money or material.

Such assistance, based on the incorporation of trees by and for the people, will cost less than large-scale reforestation on state lands. In fact, for the same amount of money, the achievement, measured either by area restored or by number of trees planted, will be much greater than if the work is carried out by the administration alone. Finally, this assistance will do more for the rural economy than could be expected from wood production alone.

Arid zones present problems of land use and of stabilization of the bases of agricultural production. Forests, trees and shrubs, as well as annual plants, have an important role to play in this connection and represent perhaps the primary means of arriving at a balanced rural economy and perhaps even the only way of avoiding degradation of the land. As elements essential to the overall stability of agriculture in these zones, trees cannot be relegated to a secondary role but must form an indispensable component of rural management, and this must be managed with and for the people within the framework of a new administrative orientation and with adequate financial and technical assistance.

REFERENCES

Eastin, I. 1996. *The role of marketing in promoting a sustainable trade in forest products in tropical regions.* Paper presented at the FAO/UPLB Workshop on Forest Products Marketing, Los Banos, Laguna

Jahnke, H.E. 1982. *Livestock production systems and livestock development in tropical Africa.* Kiel, Germany, Kieler Wissenschaftsverlag Vauk.

Mayadill, H.T., 1991, Agroforestry for Tropical Rainforests. Institute for Tropical Rain forest and Ecology. Agroforestry systems. 13(3):259-267 p.

Swindale, L.D. 1982. Distribution and use of arable soils in the semi-arid tropics. In *Managing Soils Resources—Plenary Session Papers.* Proc. 12th Int. Congress Soil Sci., 816 February 1982, New Delhi.

9

Domestication of Trees in Agroforestry Systems

The domestication of a plant species involves its manipulation and cultivation for specific uses. Through breeding and selection this process can result in rather uniform plant populations with a narrow genetic base. The cultivation of such selected varieties is the last stage in an evolutionary continuum in people-plant interactions. In this evolutionary process a progressively closer interaction between people and plant resources takes place with increasing inputs of human energy per unit of exploited land.

Three dimensions of domestication may be distinguished, i.e., domestication of the biophysical as well as the human environment and the domestication of the plant species. The recognition of these different dimensions is of particular importance with respect to the domestication of tree species, because commercial tree utilization usually starts with the exploitation of trees from the wild in natural forests. The domestication of the forest environment involves, in the first instance, the change from uncontrolled utilization of the wild tree products to their controlled exploitation. The second phase consists of the purposeful cultivation of wild trees in either a resource-enriched natural environment or in indigenous agroforestry systems such as forest gardens. The cultivation of domesticated trees in either agroforestry systems or commercial tree-crop plantations is the last stage of this domestication process.

These three stages of domestication are associated with specific socioeconomic and ecological conditions. During the first stage of domestication of the forest environment the management practices are, in first instance, directed at defining user rights. In the second stage, management practices with biological objectives are also undertaken. These focus at the manipulation of the growing conditions of the trees and of the tree's morphological characteristics. During the third stage of plant domestication the genetic constitution of the tree population is also modified. The recognition of various phases of forest and tree domestication provides for a better understanding of the different options for the production of commercial tree crops in agroforestry systems, under location- and time-specific conditions. It also illustrates that efforts to domesticate commercial forest resources should be focused both on the manipulation of morphological characteristics of a tree (its environment) and on the modification of the tree's genetic characteristics.

NATURE OF PLANT DOMESTICATION

The concept of plant domestication is interpreted differently by various scientist. Some have defined it in a relatively narrow sense as a human-induced change in the genetics of a plant, to adapt it to human agroecosystems; this process ultimately culminates in the plant's inability to survive in natural ecosystems. Others have defined plant domestication in a broader sense, as a process of naturalization of plant species towards specific human-induced growing conditions, during which an increased adoption for specific uses normally takes place. For instance, Prance describes that extraction of non-timber products from the forests may be considered as the initial phase of domestication of valuable tree species. During the domestication process, wild plants are first brought to some form of management. In a later stage of the process, the wild plants are actively cultivated. Only in its final phase does the process involve the selection and breeding of selected genotypes resulting in rather uniform plant populations with a narrow genetic base.

The concept of plant domestication can thus be interpreted in a restrictive sense as a process of changing of the biological characteristics of a species, or in a more extended sense as a process of change in plant exploitation practices, which brings with it changes in the plant's morphology and genetics, as well as in its growing environment.

Some authors have even extended the concept of domestication from the conventional biological definition to a more inclusive one, i.e., the process of increasing human-plant interactions. According to this interpretation, domestication not only involves a change in plant characteristics and the biophysical environment, it also involves adaptations in human activities with respect to the use and manipulation of valuable natural resources. The various dimensions involved in the process of domestication can be clarified by relating it not only to plant species but also to landscapes: 'A domesticated landscape is one that has been modified by humans from its highly biodiverse state to a state that may still have high biodiversity, but which contains a greater concentration of resources useful to humans'.

Such an integrated concept of domestication within the landscape seems particularly apt when we consider agroforestry. Many agroforestry systems develop as a result of the gradual modification of forests, by enriching them with useful crops. Such enrichment not only involves agricultural crops; in many cases it also concerns trees producing non-timber forest products (NTFPs). The first phase of domestication of NTFP trees involves a process of concentration of naturally occurring useful tree resources in natural forests. In subsequent phases, also, new species may be introduced into these forests. At first, this will take the form of transplanting of wildlings or seeding of wild species. But gradually selected varieties will also be introduced. As a result, the natural forests gradually change into a mosaic of managed forests and agroforestry systems.

The process of domestication can therefore be considered as an evolutionary process, from gathering to breeding, during which changes at the level of both the landscape system and the plant species occur. Concomitantly, a progressively closer interaction between the tree resources and people takes place. The domestication of NTFP trees in agroforestry systems thus involves a coevolution of human society and nature, with the combined processes of natural and cultural selection creating a great diversity of human-influenced agroforest types.

This coevolutionary process consists of various stages along a nature-culture continuum. These stages can be associated with specific socioeconomic and ecological conditions. To understand under what conditions farmers are interested in incorporating commercial NTFP trees in their land-use system, it is necessary to understand the specific characteristics of these different stages. This will allow identification of the kind of domestication practices that offer the most scope under specific socioeconomic conditions and for given agroforestry systems.

PRODUCTION OF NON-TIMBER PRODUCTS

The potential of tropical forests to provide a variety of non-timber forest products is at present widely acknowledged. Such products include a variety of food products as well as diverse commercial crops such as gums, resins, rattan. In the past it was often assumed that mainly tribal people who were engaged in traditional methods of forest resource use, such as gathering and swiddening, collected such products for their subsistence. But it is increasingly acknowledged that there exists a long history of commercial trade in non-timber forest products, coupled with an active management of these products.

When considering the production of non-timber forest products, it is therefore not appropriate to assume a dichotomy between the gathering of NTFPs from wild trees in a natural forest and the cultivation of domesticated NTFP trees in an agroforestry system. Rather, attention should be focused on the various phases of domestication and on the creative role of human culture in regulating tree resources for human use.

Agroforestry is a generic name for a variety of land-use systems in which trees are purposively combined with agricultural crops and/or livestock. With respect to NTFP trees, a wide range of agroforestry and other forest systems may be distinguished:

- *Native forests in which NTFPs are protected:* Specific areas or specific tree species in native forests that are favoured and protected because of their value for providing useful materials.

- *Resource-enriched native forests*: Native forests, either old growth or fallow vegetation, whose composition has been altered by selective protection and incidental or purposeful propagule dispersion of food and/or commercial species.
- *Reconstructed natural forests:* (Semi-) cultivated forest stands with several useful planted trees, together with tolerated or encouraged wild species of lesser value, and non-tree plants (herbs, lianas), composed of mainly wild species.
- *Mixed arboriculture*: cultivated mixed stands, almost exclusively of planted, and often domesticated, tree species.
- Interstitial trees on croplands: Either naturally regenerated and protected trees, or planted and sometimes domesticated trees scattered over agricultural fields.
- Commercial plantations with associated agroforestry practices: Plantations of domesticated tree crops that are (temporarily) intercropped with food plants or grazed by livestock.

This categorization is a first approximation of the varied and often complex agroforestry systems incorporating NTFP trees. The different categories are not discrete: gradual transformations from one category to another may take place. Many of these systems are indigenous in nature, and they have gradually evolved in response to changing conditions. Such changes may involve a variety of ecological, socioeconomic, cultural and political factors. The four most important changes that may affect agroforestry systems are:

- changed ecological conditions, such as resource depletion or land degradation
- changed technological conditions caused by the introduction of new agricultural and forestry techniques
- changed economic conditions, such as development of new markets and increased commercialization, changed demands for forest products and changed opportunities for off-farm employment
- changed sociopolitical conditions, e.g., population growth and migration, increased interaction with other ethnic groups, changed tenure conditions including gradual privatization or nationalization of forest lands, development of state organizations for forest management and rural development

These changes increase pressure on forest and tree resources. In several cases, this has resulted in deforestation and forest degradation. But in other cases, farmers have reacted by modifying their management strategies, e.g., by intensifying the cultivation of valuable tree species. For instance, management for fruit production may evolve from collecting wild fruits in the forest, through their cultivation in enriched fallows and home gardens, to fruit production in orchards.

Such changes in indigenous agroforestry systems demonstrate that many local communities have been actively engaged in domesticating the landscape as well as valuable tree species. Efforts to stimulate the further adoption of domesticated NTFP trees in agroforestry systems will be most successful if they build upon and strengthen such indigenous processes of domestication. This is possible only when a good understanding exists about the various dimensions of domestication in agroforestry systems, as well as the specific characteristics of the various phases during this process. To obtain such understanding, it is useful to consider the various dimensions of tree management in more detail.

MANAGEMENT OF TREES IN AGROFORESTRY SYSTEMS

Often, the concept of forest or agroforest management is used with respect to the conscious manipulation of the environment to promote the maintenance or productivity of forest resources. Such a definition is not complete, however; it limits management towards the implementation of certain technical activities. It does not incorporate the notion that management includes not only the execution of activities but also the process of making and effecting decisions about the use and conservation of agroforestry systems, as well as organizing the management activities. Agroforestry management can therefore better be conceptualized as referring to the total set of technical and social arrangements involved in the protection and maintenance of agroforestry resources for specific purposes, and the harvesting and distribution of the products.

Technical Management Practices

To date, most studies on domestication of NTFP trees in agroforestry systems have concentrated on the scope for either improved regeneration practices or selection and breeding of specific NTFP species. This illustrates that many researchers interpret the process of NTFP domestication as involving basically a change in biological properties of NTFP trees, rather than as a process of change in human-plant interactions. When domestication is considered in this more inclusive perspective, it becomes clear that management practices in agroforestry systems do involve more than purposeful regeneration of selected tree species. Management practices also include the stimulation of the production of the required products and the protection of valuable tree species against damaging agents.

When considering how and to what extent domestication relates to these various management practices, it is important to recognize that it is not correct to assume that planting trees is the first stage in agroforestry management. Although this is true when bare land has to be replanted, it is not the case when agroforestry systems are developed in existing natural forests. In the latter case, the first phase of management consists of protecting valuable tree species through controlled harvesting techniques. Such controlled utilization of forest products involves not only biologically oriented practices but also the definition of user rights.

If ample forest resources are present, there is no interest in forest protection or tree planting practices, and management practices may then be confined to defining use rights. But if forest resources are perceived as becoming scarcer, management may be intensified by also implementing biologically oriented practices that limit overexploitation, e.g., limiting the harvested amounts or harvesting rotationally. This management may further be intensified by the execution of practices aimed at increasing the presence or productivity of the desired species. Measures may also be developed to favour the useful species indirectly by manipulating the biophysical environment and removing less desirable competitors.

Thus, management practices consist of a group of deliberate activities for the conservation and possible enhancement of useful forest resources and the controlled utilization of those resources. These practices may be directed at influencing either the forest resource directly or its biophysical and social environment. Much variation in the kind and intensity of management practices may occur between various agroforestry systems, depending on the kind of resources being considered and on the kind of management system. Gradual transformations from one system to another may take place. During this progression, valuable tree species are gradually segregated from the natural forest and cultivated in increasingly specialized agroecosystems. Concomitantly, the vegetation structure becomes increasingly systematized, with randomly spaced trees of random age giving way to even spacing of even-aged trees. Furthermore, the propagation methods change from using seeds to using clonal material, and the location of regeneration units changes gradually from a forest environment to open-field conditions.

Organization of Forest Management

Agroforestry management involves not only carrying out resource management practices but also the process of making decisions about:

(1) the objectives of management,

(2) the kind of activities to be carried out by various persons, and

(3) the distribution of products.

In addition, management also requires the existence of a control system that ensures that the proposed activities are carried out as planned. In most studies on tree domestication no attention is given to such organizational aspects. It seems to be assumed that the production of NTFPs takes place within clearly delineated management units and under a clearly defined management authority. In many cases, however, this is not necessarily the case. In many instances agroforestry development is planned in areas that are characterized by a pluriformity with respect to land and tree tenure.

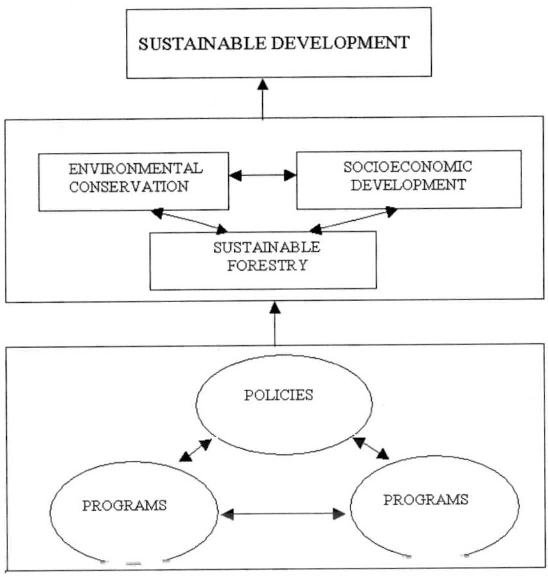

Figure 1. Sustainable Forest Management

Several agroforestry systems are managed not by private owners but by local communities who act cooperatively in managing agroforestry systems as a common resource. Such agroforestry systems are subject to individual use, but not to private ownership. The utilization of these resources is governed by a set of regulations on independent user rights of members of a specific user group. For such common property forest management regimes to function properly, there should exist a local forest management institution with the following characteristics:

- A structure for group members to make decisions on the required resource management practices
- Group control over the behaviour of the group members, which ensures that the planned management practices are carried out
- Control over the distribution of collected forest products
- Ability to exclude outsiders

The common property management regimes are mostly characterized by regulations defining user rights and measures for controlling overexploitation. Often no measures for stimulating biological production are included. But when the management is further intensified by investing labour to increase productivity, the tenure regime often shifts from common property rights to private ownership rights.

EVOLUTIONARY STAGES OF TREE DOMESTICATION

According to the basic principles of agroforestry management, the process of domestication of valuable tree species in agroforestry systems may be characterized as a general evolution from extraction of products from the natural forests under common property regimes to the cultivation of domesticated tree crops on private lands. This evolutionary process thus involves both technical and ecological changes as well as institutional changes. Although examples of different stages of this evolutionary process have been described by various authors, to date the systematic comparative analysis of the various stages of domestication has received little attention.

In an attempt to improve understanding about the evolution in forest-people interactions, Wiersum (submitted) has developed a model in which various forest exploitation and management activities are arranged along a gradient of evolutionary stages in people-forest relations. This model was based on a model by Harris in which various stages in exploitation of agricultural crops were arranged sequentially along a gradient of increasing input of human energy per unit of exploited land. On the basis of the principles of forest management, three major categories of forest management practices may be discerned. These categories can be conceived as representing progressive phases with respect to the input of human energy per unit of exploited forest, i.e.-

- controlled utilization of forest resources
- protection and maintenance of forest stands
- purposeful propagation of valuable forest components through the regeneration of wild or domesticated trees

Consequently, three major thresholds may be postulated between the various phases of forest-people interactions. The first is between uncontrolled and controlled procurement of wild tree products in the natural forests. As indicated above, the control of forest utilization primarily involves the definition and control of use rights; this requires social transaction costs with respect, for example, to time spent on mobilization, decision making and control. Control measures with a biological objective are developed in the second instance. The second threshold is between controlled procurement of wild products and purposeful regeneration of valuable tree species. And the third threshold is between the cultivation of wild and domesticated tree species; the domesticated trees may be propagated through genotype or phenotype variants.

A gradual transformation of the natural ecosystem into an agroecosystem occurs concomitantly with this increasing input of human energy per unit of exploited forest land. In addition, the human intervention in the reproductive biology of tree species intensifies. This process of progressively closer interaction between people and forest resources is associated with various socioeconomic trends. In the first place, the socioeconomic conditions relating to forest utilization change: increasing sedentarization, increasing population density, and a gradual shift from a subsistence economy to commercialization. In the second place, the complexity of the indigenous rules and regulations change, with common property rights gradually becoming changed into private land and tree tenure rights. Thus, the forests and trees are incorporated in an increasingly more complex cultural environment.

This model should not be regarded as an explanatory model indicating unidirectional and deterministic trends, in which the various phases represent preordained steps on a ladder of increasingly 'advanced' stages of general societal development. In effect, in many areas different agroforestry types may coexist, with each type occupying a specific landscape or tenurial niche. As agroforestry systems are mostly a component of an integrated farming system, the evolutionary trends are anything but straightforward.

Depending on their role in the local farming systems the management practices may either be intensified or deintensified in response to agricultural intensification or to more general changes in socioeconomic conditions. Nonetheless, the model may assist in clarifying the various stages in the process of the domestication of valuable tree species in agroforestry systems. It illustrates the various phases in the process of domestication of forest resources and identifies the general relations between tree management practices and ecological and institutional conditions. This information can assist in assessing what kind of activities might be considered when stimulating further domestication of NTFP trees in various categories of agroforestry systems.

DOMESTICATION AND COMMERCIALIZATION OF NTFPS

Domestication of forest species for purposes of commercial cultivation seems to have bright prospects. Forestry definitively needs to find an answer to the exhaustion of wild resources while in the same time rationalising the production of marketable NTFPs. There is great demand for new crops and new markets.

Sustainable development has to mitigate the effects of deforestation by increasing the planting of trees on cleared lands, and tropical farmers have to find substitutes for the natural resources lost through deforestation. However, domestication is neither a fast nor an easy process. It took centuries of farmers' work to develop a range of food crop varieties and decades of scientific research to create productive clones of industrial tree crops. Can foresters or farmers wait another 50 years to tame wild rattans, or before they efficiently produce natural resins or marketable 'forest fruits' from improved and domesticated trees? Can we wait for agroforestry research to spend several more decades testing efficient tree-crop associations for forest domesticates and making them acceptable to smallholder farmers?

Modern plant domestication relies on relatively sophisticated techniques. But it should not be restricted to only a set of techniques. Domestication is also, and primarily, a strategy. Domestication techniques are selected on the basis of economic choices and by the initial preference for a given cropping system within a given farming system. Domestication, this latter aspect is often neglected, probably due to the historical background of the present scientific community.

Modern agricultural and forestry science evolved in Europe, in a pastoral and cereal-growing civilisation, which influenced most of the present agricultural paradigms and agroecosystem models. Our perceptions of domestication and agricultural development are deeply influenced by a tacit preference for cereal-based models, which may not be the most useful for the domestication and development of tropical forest resources. The design of the agroecosystem in which domesticated species are grown is as essential as the choice for particular plant selection, breeding and reproduction techniques. This is particularly important when switching from annual crops to trees and from fields to forests.

Most 'modern' agroecosystems in the world relate to a single agricultural model: intensive, highly specialised stands of homogeneous crops. However, other available models exist, which should be examined under the new perspective of domestication of tropical forest species. Especially important among these are the native farming systems in the tropics, for which there is academic information already. This presentation will concentrate on the implications that ecological models in such native systems might have for the design of novel domestication strategies for NTFPs, particularly in agroforestry.

An 'ecosystem' perspective to the domestication of forest species could change the approach chosen and lead to changes in techniques and processes. It is essential to understand that the importance of an ecosystem approach to domestication goes far beyond biological or technical considerations. It also has obvious sociopolitical and institutional dimensions.

The common preference of governing elites and scientists for the plantation model of domestication for commercial forest resources in the tropics did not only change the face of forest landscapes and national economies. It also deeply affected forest communities and their socioeconomic life. It is essential to analyse in the context of NTFP domestication, this indirect link between the choice of a particular domestication strategy and the fate of local populations.

Domestication is part of a resource appropriation process, and resource appropriation by a powerful fraction of the population might lead to dispossession of the weaker fraction. This aspect of domestication will be examined in the light of the Indonesian history in forest resource management.

Food Crop Domestication

The process of domestication and cultivation of food crops has followed two totally divergent routes: 'agriculture' and 'horticulture', taken in their etymological sense-the cultivation of 'ager' (openfields) and the cultivation of 'hortus' (gardens). Several ethnobotanists and anthropologists have suggested that relating these 'grain' and 'garden' models to existing cropping systems will help.

'Agriculture' refers to the 'grain model' developed for cereal domestication in ancient Mesopotamia and around the Mediterranean, and also in ancient rice civilisations from India to China. 'Ager' conquered the forest and is the central platform for domestication and the 'home' of domesticates. It represents a highly specialised and artificially uniform openfield, devoted to grain crops, largely disassociated from the pre-existing environment. It involves a massive population of genetically homogenous plant cultivars.

The grain model relies on highly specialised and segregated technical knowledge relating to monocultures as well as to intensive use of chemical and mechanical inputs. These are associated with very high energy consumption and minimal labour inputs at the field level. Genetic specialisation and manipulation culminate in the widespread use of hybrids, which totally depend on people for their production and regeneration. The grain model perfectly relates to the productionist mentality of modern agriculture. Initially devised for food crops, it has deeply influenced modern commercial tree cropping: both tropical plantation agriculture as devised by European colonialists and commercial fruit culture in temperate regions. Both situations replicate a modern corn field biologically and technically, as well as in ideology. 'Horticulture' refers to the management of 'hortus', the garden, which can be characterised by high plant diversity, including tuberous perennials and trees, in somewhat chaotic combinations and configurations.

The garden model involves diverse production, from many food crops. Domestication in the hortus, whether in temperate or tropical regions, has operated through the treatment of plants as individuals. It developed countless cultivars of fruits, tubers and vegetables. In the tropics, 'hortus' can be a swidden, an anthropogenic forest, or a homegarden. Management in the garden model plays intensively on ecosystem interfaces and fully benefits from natural vegetation dynamics. Devised for multipurpose production like indigenous agriculture in the tropics, as well as for an optimum management of ecological and economic risks, the garden does not comply with the urgent need for 'productivity' in agriculture. Temperate gardens are nowadays devoted mostly to production for home consumption or to leisure, whereas their tropical relatives are neglected or even denigrated by official agricultural services. In the one-tracked mind that characterises modern rural development, the grain model is considered as the only model that is valuable for efficient agricultural production.

Agroforestry research and extension represents the only consistent scientific framework through which the hortus model could be further developed. But this requires innovative approaches, especially as far as domestication and cultivation models are concerned. The openfield preference has often led to reductionist approaches to domestication, which we have to re-examine in the light of agroforestry development. Does domestication unavoidably come down to adapting a wild species to grow in industrial plantations? Are tropical gardens confined to the production of, at best, semi-domesticates? What new perspectives can agroforestry bring to domestication?

Domestication Models

Domestication and cultivation have long been the prerogative of agriculture, while worldwide, forests remained the domain of hunting and gathering in support of agriculture. Until roughly the beginning of the 19th century, tropical and temperate forests were mainly managed for integrated, multipurpose use. These practices more or less maintained forest composition, providing for grazing, hunting, gathering, and wood or timber production.

Some species selection did happen, creating useful genotypes that could be substituted for wild ones in niches within the pre-existing biotic community. Among other examples are productive chestnuts and sweet acorn varieties in the Mediterranean countries, and peach palm and brazil nuts in the Amazon basin. Rationalisation of forest production and forest culture did not really appear before the industrial era. This coincided with more focused demands for specific forest products: fuel for metallurgy, timber and pulp.

The development of scientific forestry induced a strict partition between agricultural and forest development. Productivism applied to forestry created new models of intensive wood production, inspired from the grain model in agriculture, with the multiplication of highly homogeneous, specialised, and productive plantations of pine in the north and of *Acacia* and *Eucalyptus* in the tropics. Today's commercial forestry in the tropics, dominated by the imperatives of wood supply, follows a bipolar model:

products are either extracted from natural, more or less managed natural stands, or cultivated in highly specialised plantations.

The monoculture preference stands as the recognised option for domestication strategies for timber trees. Domestication and cultivation techniques devised by forestry research for tree species follow the rules of specialisation, uniformity and intensification that have proven efficient for grain crops, simplifying the structure and the function of the cultivated 'forest' to the extreme. But how far does this model extend to NTFPs?

Domestication and Cultivation of Tropical NTFPs

Domestication of non-timber forest products in the tropics developed under three different situations.

Indigenous Practices

The history of management, selection and cultivation of useful forest species by indigenous people for subsistence purposes is as old as that of humankind's use of forest ecosystems. In terms of domestication, the process of interaction between farmers and forests gave rise to a whole range of modified tree varieties, which most authors classify as 'managed species' or 'semi-domesticates'.

However, scientists commonly recognise that true domestication occurred through the transfer of edible forest species-primarily fruits and nuts-to agriculture through various types of garden. These forest domesticates are presently considered as true agricultural or horticultural crops. Though not often acknowledged in the literature, commercial gathering of forest products by local people also went along with plant selection, integrated management and even cultivation.

This simple domestication developed interesting models of true 'forest culture', that relate to the hortus model. These models have unfortunately remained neglected are almost never included in development programmes, even in agroforestry.

Incorporating forest resources in farming systems is not a new practice in the tropics; various traditional forms of agroforestry have developed around NTFPs, constituting the very basis of indigenous agricultures in tropical America, Africa and Southeast Asia. Beside fruit-dominated gardens, indigenous farmers have integrated true forest culture into farm lands, and these forms are usually closely associated with shifting cultivation. These 'cultivated forests', which often complement subsistence food cropping in annual fields, are established after total removal of the original vegetation, and they constitute complex, tree-based agroforestry systems, which fully deserve the name of 'agroforests'. Why have they been overlooked in agroforestry research? Perhaps it is a consequence of their appearance: they do not look like cultivated ecosystems and have mainly been mistaken for natural forests.

Many true 'agroforests' in Indonesia have evolved around fruit and nut trees. But there are other systems that have evolved from former extractive practices in natural forests, through the deliberate incorporation of non-food forest tree species that farmers have cultivated for products to be marketed in international trade.

Among the commercial forest products, cinnamon (*Cinnamomum burmanii*) was probably the first to have been incorporated into indigenous agricultural systems. In the central highlands of Sumatra, indigenous stands have been established for more than two centuries. Some form specialised, homogeneous gardens, but others, established on steep slopes, associate cinnamon trees as an understorey with higher canopy trees grown for fruits or timber. Similarly, benzoin (*Styrax*) is known to be managed as a fallow crop in what represents a true rotational agroforestry system in Laos. Such rotational systems were also mentioned in North Sumatra as long ago as the 18th century. However, other cultivation practices also developed through more complex and permanent agroforestry systems, which associate benzoin trees in a mix of useful timber and fruit trees.

In western Borneo, swiddeners have, for at least 150 years, established highly diversified tree gardens that integrate oil-producing dipterocarps together with tens of other fruit and nut species as well as rattans, latex-producing trees and timber species. In Central and East Kalimantan, rattan, which has traditionally formed the bulk of trade in forest products, has been incorporated into shifting cultivation systems for more than 100 years. Rattan gardens mix the cultivated palms with planted fruit and timber trees, as well as with numerous other useful species that have established spontaneously.

A century ago in the south of Sumatra, swidden farmers started cultivating damar trees for resin production and have established more than 20 000 hectares of complex forest-like agroforests, associating damar with numerous other fruit and timber tree species. Native rubber trees also happened to have been planted in complex gardens, but rubber agroforestry really developed with the incorporation of the para rubber tree in local swidden systems at the beginning of this century.

The Amazon rubber tree found its ecological niche in complex tree gardens in Southeast Asia, where it is grown with numerous other species, either planted or spontaneously established. It soon replaced native rubbers in the economic niche of the local swidden farmers.

All these agroforests result from farmers' needs and their deliberate choice to improve production and control, or sometimes to protect or even restore useful forest resources. They all present important common features: (1) most of them concern true old-growth forest species, not fast-growing pioneers, (2) they have all evolved from swiddens, through the systematic introduction of trees in cleared lands, (3) most often, they start as specialised plantations that evolve into a permanent mixed stand of planted tree crops and useful spontaneous resources, and (4) they exhibit forest-type structure, including a predominance of large trees, a multilayered vertical configuration and a closed-cover canopy.

Some of these indigenous systems, like the dipterocarp agroforests, hold structural as well as functional characteristics typical of a primary forest ecosystem, with the predominance of big trees, a high species richness, a high ecological complexity, and a closed nutrient cycle. Others, like the rubber agroforests that cover the lowlands of Sumatra and Kalimantan, are more like secondary forests, with dense stands of smaller trees and a rapid turnover of species.

These agroforests combine important income-generating strategies based on forest resources and diversified subsistence strategies, but they are not isolated management units: they always complement other agricultural activities, such as food cropping in open fields. Lastly, they rely on local representation and knowledge systems evolved from former forest traditions; they are maintained by simple techniques and integrated practices and are controlled by a well-defined social and tenurial system, which includes rights as well as duties.

Colonial Interventions

Tropical NTFPs acquired a new dimension during the expansion of the colonial era. This started with the high demand for spices (nutmeg, cinnamon, vanilla), coffee or cocoa, which created a new, but highly specialised commercial perspective. These cash crops boomed with the rise of the industrial era, which needed new products, such as latexes and resins. For some products, a demand boom, the need for increased control over the resource or over product quality, led to domestication through intensive cultivation.

The process of domestication and cultivation developed by the colonial plantation managers induced a true movement of disassociation between forest resources and forest ecosystems. Most of the new economic forest resources were transferred to plantation agriculture, through vast estates that reached their peak in the second half of the 19th century and the early 20th century, with oil palm, rubber, cinchona, coffee, tea and cocoa. In transferring forest resources to agriculture, the colonial plantations did not attempt to innovate; they followed the grain model, creating huge areas of specialised, artificial and productive tree monocultures.

New Trends

The commercial interest in NTFP declined after World War II, because of the fall of colonial powers and the rise of petrochemical industries, which developed substitutes for natural products. However, a new focus on NTFP exploitation has recently emerged. NTFP extraction is now considered a promising alternative to timber extraction in natural forest management.

The justification for this revived interest is not entirely based on economics, although new markets for natural products have emerged, such as phytochemicals in pharmaceutical industries or wild substances in food industries, that have triggered economic interest in NTFP. The other justifications have been ecological and sociopolitical.

Ecological interest has focused on reduced disturbance to the forest ecosystem *vis à vis* timber extraction practices-while sociopolitical interests have centred around the promotion of new development models for indigenous forest people and the promotion of 'fair trade' for natural products. For some of the most coveted products, development through domestication and commer-cialisation tends to emerge as a strategy competitive with extraction from natural forests. Thus, despite the debate about developing new and better models for the domestication and cultivation of useful tree species in farm lands, there is a good probability that the conventional monocultures will persist. This is already obvious for rattan production in Indonesia and Malaysia, where specialised plantations have been established under the strict control of forestry services, and for the Brazil nut in Brazil, where private investors have established large plantations that severely compete with smallholder farmers and extractivists activities.

Consequences of NTFP Domestication

The era of colonial trade and management had two major consequences for forest people: the displacement of their effective control over the collection and trade of forest resources and passage of the resources either to the governing elite or to private colonial entrepreneurs. This resulted in the local forest communities being overlooked and in the consequent transfer of the commercial benefits to private and commercial planters.

This 'dispossession process' continued through most post-colonial governments and is still common today when a traditional forest resource encounters a commercial boom. This abuse of native rights commonly starts with restrictions over harvesting practices, develops with the attribution to acknowledge traders of monopolistic rights for gathering, and culminates with plantation development. This process has political support and is a practical consequence of commercial interests. The history of natural rubbers in the Amazon and Southeast Asia perfectly illustrates this dispossession process, which transferred control over the resource-from native collectors to powerful traders and then to planters. Caoutchouc was traditionally collected and used by Amerindians in the Amazon Basin. In Southeast Asia, especially Sumatra and Borneo, other wild rubbers (*Ficus elastica* and *Willughbeia* spp.) were harvested and traded by local swiddeners before the 17th century.

Commercial interest for wild rubbers in Europe and the United States boomed in the second half of the 19th century, but until 1900, both elastic and non-elastic rubbers remained supplied exclusively by wild species through latex extraction. As prices rose, the control of the rubber areas in Amazonia became concentrated in the hands of the political-economic elite, who 'exerted an absolute rule over native rubber tappers'.

At the same time in Indonesia, the Dutch colonial government progressively restricted exploitation by local tappers, first through imposing a licence to tap the trees, then through granting all tapping rights to foreign concessionaires. Another period of dispossession of native tappers started with the cultivation of the Amazon rubber tree in Southeast Asia in 1877. By 1913, the supremacy of the wild caoutchouc came to an end and the local tappers in the Amazon virtually stopped working. Cultivated rubber production in the Far East, through large estates, had captured the market.

Such examples are numerous, starting from the Dutch VOC taking possession of local nutmeg production and trade in the Moluccas in 1621 to the present Indonesian seizure of the birds' nests caves of the local Punan in East Kalimantan, who had owned and managed them sustainably for centuries. To get full control of the nutmeg trade, the Dutch established specialised nutmeg plantations worked by imported slaves while destroying nutmeg trees in surrounding islands. Nobody except VOC was allowed to grow and trade nutmeg any more.

Prospects for NTEP Domestication

Forestry in the tropics is still looking for technical and economic as well as socially accepted models. Can it escape from the bipolar model extraction in natural forests on one hand and monospecific specialised plantations on the other? Can domestication of forest trees help to alleviate this situation, or will it contribute to increase the present movement of segregation between forest and agricultural development? Who will benefit from domestication of NTFP: foresters, planters, or smallholder farmers? How can rural development efficiently and sustainably incorporate the so-called minor forest resources into farm lands?

Agroforestry is often cited as the most favourable means of providing positive answers to such questions. But what type of agroforestry is the best for NTFP development? Is agroforestry research able to fully integrate and develop the potential offered by these 'new' crops? Can it generate 'new' models of agroecosystems, combining both agricultural and forest qualities in an economic as well as ecological perspective?

Most agroforestry research until now has concentrated on simple associations, trying to introduce a single tree or shrub species into former grain-model systems. Most agroforestry research has promoted fast-growing trees or shrubs, not old-growth forest species. Agroforestry has been more 'agro' than 'forestry' oriented, more crop than tree based. When dealing with NTFP domestication and cultivation, agroforestry research definitely has to innovate. It has to give new preference to tree-based systems and to look for new ecosystem models.

Ecosystem Domestication

The current concept of tree domestication appears poorly adapted to characterise and critically analyse most of the above-mentioned examples of integrating forest resources into agricultural farming systems. 'Domestication' usually focuses more on selection and propagation techniques, making too little reference, if any, to the concepts and strategies developed for the integration of wild resources direct into the farming system. This last point is nevertheless essential when dealing with the domestication of wild forest species.

The preeminence of the grain model has obscured analyses from other perspectives. Is the conventional model of domestication the best choice for trees that have evolved in a highly diverse and structurally complex environment? Usually, domestication has intentionally disassociated the resource from its natural habitat. Transfer of the candidate species to an artificially prepared environment has been seen as essential, to allow increased human control of the plant, as well as to induce and efficiently select useful genetic variations. But should the artificial cultivated environment be fundamentally different from the natural one? Although a very artificial environment has proven its efficiency in cereal domestication, as well as in colonial tree-crops development and modern forestry plantations for wood production, there is seemingly no satisfying theoretical answer for NTFPs.

The domestication strategy exhibited in the Indonesian agroforest examples partly relies on conventional plant species domestication techniques, but it does not involve crop management in highly specialised stands, which are quite different from the original conditions in which the wild species had evolved. Nor does it involve a major modification of the structural and biological features of the tree species, in which trees are selected to allow their adaptation to homogenous monocultural conditions. Rather, the agroforest model relies on an artificially induced reconstitution of a true forest-like ecosystem, simulating the basic principles of a natural silvigenetic succession, which allows the selected species to establish, grow and reproduce as in their original habitat. Establishing an agroforest is conceived as a specialised tree-planting process aimed at controlling and concentrating the selected forest resource; but this process is achieved through the integration of the resource with natural vegetation and through several successional stages that lead to the gradual reconstruction of a diversified forest structure.

The forest tree seedlings are introduced into the forest clearing with other short- or medium-cycle crops (e.g., rainfed paddy, vegetables, coffee bushes, pepper vines) and receive the care given to the crops. After the abandonment of these food crops as the tree canopy closes, the planted trees are strong enough to grow along with secondary vegetation and overcome competition from pioneers. The subsequent tree fallow then freely develops with little damage to planted trees. The structure of the agroforest becomes more complex over the years, as the consequence of a particular form of management that maximises the use of natural production and reproduction processes in order to minimise the rarest economic factor: labour.

In a maturing agroforest, plant species regenerating from the neighbouring forests, through natural dispersion, can establish while forest animals find shelter and feed. Through selection, farmers favour economic resources, but non-economic resources are allowed to reproduce. And after several decades of such a balance between free-functioning and integrated selection, the mature phase of the agroforest resembles a natural forest more than a conventional tree plantation. This ecosystem-analogy strategy has proved efficient for quick acclimation of a true forest tree species. Forest farmers in Sumatra have succeeded in what most foresters dream about, but have failed to achieve: the establishment, maintenance, and regeneration of a healthy dipterocarp plantation at low cost and on a huge scale. This is a unique example for the whole forestry world.

Dipterocarp agroforests rely on selected and planted forest trees; they exhibit high-density stands and good productivity, but they are also characterised by good ecological sustainability, low-cost establishment and easy regeneration over years. This is quite uncommon in conventional plantation forestry. The agroforest domestication process allows the maintenance of biological diversity in the qualities of the tree, as it does not focus on the selection of single-purpose varieties: the multipurpose dimension of the wild species is not lost through domestication, as usually happens in the conventional process. Low-canopy varieties of durian (*Durio zibethinus* Murr.) or rambutan (*Nephelium lappaceum* L.) selected by plant breeders produce nothing else but fruits, but damar or benzoin trees domesticated for resin production by local farmers in Sumatra are still good timber producers.

This multipurpose dimension is also maintained at the ecosystem level. In the plantation model, what is not the crop is a weed. In the agroforest, self-established species are integrated as economic and ecologically beneficial resources or kept as potentially useful species. But the most original point is how natural biological processes are utilised to support the artificial domestication and cultivation process.

In the agroforest, natural vegetation dynamics are channeled, first to speed up and secure the integration of slow-growing trees in the cultivated system, then to maintain a continuous balance between obsolescence and the regeneration of the cultivated stand. Beside domesticating forest species, the agroforest allows the restoration of integral biological and ecological processes, which determine the overall survival and success of the cultivated ecosystem.

These natural processes schematically replace the high technology and energy inputs of forest plantations. Here, it is the agroecosystem that adapts to the plant characteristics. In this sense, agroforests constitute an original attempt of 'ecosystem domestication', through the full utilisation of natural ecosystem dynamics to the benefit of a selected, artificially established population of trees.

Through this, the agroforest domestication strategy proves to be successful in assimilating the problems associated with the long-term management of forest tree species-or vines, as in the case of rattan. Long-term maintenance and renewal of forest plantations is technically difficult and is invariably costly.

Theagroforest not only achieves a simple transfer of forest resources and structures. It also guarantees the renewability of these resources and structures, and of the related economic returns. However, the agroforest strategy, empirically devised by swidden farmers all over the Indonesian archipelago, bears important weaknesses that could be solved by integrated research.

The population of cultivated trees in the agroforest is usually genetically rather diverse, which affects the productivity of the crop. Though farmers do select the best producing individuals for reproduction, they cannot reliably capture genetic variation. Vegetative reproduction methods remain simple, and sanitary control in nurseries could be improved.

Technical research aimed at these weaknesses could reorient NTFP domestication for agroforestry in two ways. First would be the development of improved plant material specially designed for a complex, forest-like environment, rather than for conventional monocultural plantation conditions.

Plant selection and breeding could be aimed at taking advantage of the 'forest' characteristic of the species for both ecological and economic benefits and adapting them to farmers technical, as well as energetic standards. The second direction for research would be to test high-yielding plant varieties-wild or improved-in agroforest conditions, as is being done by ICRAF in jungle rubber agroforests. This would expand the agroforest model to new areas and improved plant material into existing systems.

Agroforest Domestication and Empowerment of Farmers

The analysis of the agroforest domestication process should not be restricted to its technical or ecological aspects. While the transfer of the wild resources of nature to the cultivated lands of agriculture is an essential process-capturing variation in natural genetic characteristics, increasing population density, stimulating cross breeding, or escaping from natural competitors and pests-it must always integrate these with major economic and sociopolitical or policy implications. This could, in future NTFPs business, enable smallholders to do more than extract products from wild ecosystems, as they do now.

By switching from the management of wild resources in traditional extractive systems to their adoption as new crops in farming systems, farmers often aim at maintaining or reestablishing their traditional authority over the forest resource base. This is obviously important when, because of overexploitation or deforestation, wild economic resources are vanishing from natural forests, or when commercial demand increases. But it might also be an important option for native farmers when politically induced dispossession threatens their livelihood.

In most ideologies and political regimes of tropical countries, agriculture secures social and legal rights over land or natural resources for smallholders better than does forestry. Integrated forest management, as empirically conceived by indigenous forest tribes all over the planet, however sustainable or profitable it is, has never been seriously considered by the governing elites and their technical councils. This is reflected in the widespread lack of legal recognition of native rights and traditional property regimes concerning forest lands and resources in tropical countries. To gain official support, native resource management systems have to evolve in a way that complies with the conventional models.

Domestication and plantation are important steps in this process; transferring wild resources to cultivated lands is both a symbolic and a political act of appropriation. But beyond this conceptual aspect lies the legal context of appropriation. Most forest lands in the tropics, and the resources they contain, are under state control; they are 'public goods'. This usually prevents any evolution towards 'privatisation' and facilitates tacit 'appropriation' of profitable forest resources, traditionally controlled by indigenous people, by those who are close to power.

Private property-for either collective or individual owners-is more readily acknowledged and expected on agricultural land. Thus, cutting the forest and planting trees might be enough to secure, if not property rights, at least the right to claim for such rights. In Indonesia, establishing agroforests has often been a major strategy for land and resource appropriation; establishing production structures and property rights that will be transmitted to further generations is an essential aspect in this particular domestication and cultivation process. Farmers in Sumatra initially planted damar-producing dipterocarp trees, in response to the depletion of wild damar trees and the need to establish a profitable forest-based economy. But as their relations with forest authorities deteriorated, the establishment of agroforests became a strategy for legal resource appropriation.

Thus, presently agroforests are also established as a claim against the closure of forest lands and resources to local communities. Through domestication and tree growing, farmers claim that they have purposefully restored and protected not only damar but the entire forest resource, in the middle of agricultural territory, upon which they hold a firmer control and rights. 'Global forest resource' appropriation is an essential component of the agroforest domestication strategy. Specialised plantations might secure the appropriation of a given forest resource, but the agroforest strategy goes far beyond that: as it recreates forest structures, it allows the restoration of the landscape in a form that conforms better to the farmers' rights and interests.

In Indonesia, therefore, the relationship of local populations with forest resources is now more closely associated with one or more types of agroforest than with the natural forest. This allows them to maintain an economy and an associated lifestyle that remain in continuity with their forest culture, from which the agroforest directly evolved; but it places it firmly in an agricultural context. The agroforest, therefore, clearly opens the way for other novel models of improved resource management in forest lands throughout the tropics.

The choice for a particular domestication and cultivation strategy is as important as the transfer of forest resources to farmlands, in determining who holds the authority over a particular resource. Domestication and cultivation integrate technical knowledge and capital investments as well as technical, labour or energy inputs; all of these may be inaccessible to smallholder farmers.

Colonial plantations have clearly demonstrated how domestication can adversely affect indigenous 'managers' of NTFPs. Will indigenous farmers be similarly spoiled when NTFP domesticates are made available by research institutes through markets or credit schemes? This seems likely when capital-intensive processes of crop establishment and maintenance lie far beyond smallholders' financial and technical capacities, and when the high productivity from plantations leads to a fall in prices of natural products and to the economic collapse of any business collecting those products.

The domestication of NTFP through these sophisticated techniques and modern knowledge might only intensify the exclusion of smallholders from the management of forest resources. In contrast, however, the domestication of NTFPs through the agroforest strategy will probably be better integrated into indigenous populations, as it relies on simple techniques, is based on local knowledge shared by every farmer, and does not imply high energy inputs. In the process of NTFP domestication, the evolution of rubber cultivation in Indonesia illustrates how plantation and agroforest development can have totally divergent effects on smallholders.

Technical and financial constraints associated with the plantation model put rubber cultivation out of the reach of smallholders. Thus, intensive rubber cultivation in colonial estates led to the exclusion of native tappers, both in Amazonia and Southeast Asia. However, when swidden cultivators in Sumatra and Borneo adopted the Para rubber plantation techniques to their production system by planting rubber trees in their swiddens, the trees grew with the fallow vegetation and soon evolved into a forest-like rubber garden.

The trees were tapped if the prices appeared interesting, and they created an agroforest system that was much less demanding in labour and technical inputs than the current estate model. This production system soon became much more competitive than the estate plantations, and since 1945 it has gained the largest share in Indonesian rubber production. Through the rubber agroforests, former indigenous rubber collectors, evicted by the plantation owners, regained their place in the rubber trade and their share in the benefits of rubber development.

NTFP Domestication: Economic Conserns

Domestication should not be disassociated from the global economic strategy of farmers. Future NTFPs plantations, whatever their model, will be part of lands claimed and developed through agricultural techniques. They will be integrated into agricultural territories and agricultural production systems. They will support local agricultural economy. The advantages of the agroforest model did not, until now, succeed in reliably capturing the trade of products from highly commercial species.

Short-term benefits are usually much higher when trees are grown in systems conforming to the plantation model. But so too are the risks and the energy consumption, with all its global ecological and economic consequences. The agroforest model, which by contrast emphasizes economic and ecological sustainability, should thus be evaluated in terms of long-term productivity, energy efficiency and economic security. It may then prove to be much more 'productive' and 'profitable' than the currently accepted models.

Agroforests, in contrast to tree crop estates, allow the maintenance of numerous tree resources to grow together and to diversify the farmer's income. If encompassed in the framework of agricultural strategies, agroforest development represents a process of forest conversion that does not go along with economic reductionism and that does not irreversibly close the economic potentialities formerly linked to the presence of natural forest. On the contrary, through the restoration of biodiversity in the agroforest, farmers maintain a whole range of economic choices for both the present and the future. Maintaining these options appears indispensable in view of the need for sustainable development. This multipurpose aspect must be kept in mind if systematic research on the domestication of NTFPs for agroforestry systems is to be carried out.

An important aspect to consider in this respect is the potential for timber production. Timber will probably become a strategic commodity for farmers in the near future, with potential benefits that might be much higher than those provided by NTFPs. Many NTFP species also have good timber, but species domestication options, and the agroecosystem design, will unavoidably influence the capacity of the candidate forest species to produce quality timber. Investing in NTFP plantings will unavoidably lead farmers to some degree of specialisation for a given product. However, opting for 'multipurposeness' in domestication and keeping in mind other potential forms of production, at both the species and the agroecosystem level, will help to avoid the irreversibility of future economic and ecological choices by smallholders.

REFERENCES

Coates A. 1987. *Commerce in rubber: the first 250 years*. Oxford University Press, Singapore. 380 p.

Dunn F.L. 1975. *Rain-forest collectors and traders: a study of resource utilization in modern and ancient Malaya*. Monographs of the Malaysian Branch of the Royal Asiatic Society 5. Royal Asiatic Society, Kuala Lumpur. 151 p.

Geertz C. 1966. *Agricultural involution. the process of ecological change in Indonesia*. University of California Press, Berkeley. 176 p.

Leakey R.R.B. & Newton A.C., eds. 1994. *Tropical trees: the potential for domestication and the rebuilding of forest resources*. HMSO, London. 284 p.

Narr K.L. 1956. Early food producing populations. In: *man's role in changing the face of the earth*, ed. W.L. Thomas. Chicago.

Pelzer K.J. 1945. *Pioneer settlement in the Asiatic tropics*. International Secretariat Institute of Pacific Relations, New York. 290 p.

Purseglove J.W. 1974. *Tropical crops*. Longman Group Limited, Harlow. 719 p.

Sauer C.O. 1952. *Agriculture origins and dispersals*. American Geographical Society, New York. 175 p.

10

Agroforestry and Tenure Issues

Many people today have great expectations for agroforestry, some of which would seem to be justified on technological grounds. However, if current efforts to understand, develop and disseminate agroforestry technology are to have any hope of meeting even a reasonable proportion of current expectations, its deployment, as a newly organised branch of applied science, must take place with a clearer than usual view of the human context of supposed land-use improvements.

The interactions between agroforestry and tenure issues are basically of two types: first, tenure factors may pose constraints to the realisation of the potential ecological and socio-economic benefits of agroforestry in many land-use systems; and second, agroforestry may offer ways of resolving or mitigating some existing tenure problems. Without doubt, tenure issues are far more varied and complex than are reflected here. However, attention is focused on some of the major changes in tenure that arise in conjunction with the main developmental trends in tropical land use. These changes are then viewed in ecological and evolutionary perspectives.

From a project standpoint there are two fundamental ways of arriving at agroforestry: by integrating trees into farming systems or by integrating farmers into forests. Appropriately selected woody components may contribute to both the productivity and sustainability of farming systems on marginal land in several ways: by enhancing the production of organic matter; by maintaining soil fertility; by reducing erosion; by conserving water; and, by creating a more favourable microclimate for associated crops and livestock.

These "service roles" are above and beyond the direct "production roles" trees can also play in supplying food, fodder, fuelwood, building materials and other raw materials for rural industries. In traditional land-use practices, agroforestry is also important in maximising and diversifying the productivity of even highly fertile lands. Intensive agroforestry systems are most commonly found in areas with a long history of population pressure, indicating their general efficiency as a land-use system. Whether on marginal or high-potential lands, diversified agroforestry systems may be the most appropriate form of land use where land tenure constraints, lack of marketing infrastructure or an unfavourable political economy make it imperative for small landholders, in trying to reduce risks, to try to satisfy most of their basic needs directly from the land resources under their control. All tropical land-use systems exhibit varying degrees of "leakiness" with respect to the cycling of nutrients held in the soil-vegetation complex, although systems such as irrigated rice paddies, permanent tree crops and forests are inherently more sustainable than others. It is a fundamental contention of agroforestry that trees have good prospects for plugging many of the holes in tropical farming systems.

The degree of "infilling" can vary from slight (limited interstitial planting) to virtually complete. Essentially, the decision as to how many and which kind of trees it is profitable to add to the existing pattern of land use depends on what useful niches for trees can be identified. An agroforestry "niche" in this sense has three components: a functional role within the land-use system; a place within the landscape; and a time within the life cycle of a particular land-use system.

Although many of the recent research thrusts in agroforestry have been directed toward the integration of trees into farming systems, agroforestry also has a role to play in the preservation of forests and the improvement of forest management systems. By providing farmers with a means of producing fuelwood, timber, building poles and other forest products on farmland, agroforestry can significantly reduce the demand on forests and natural woodlands.

By doing this in ways that enhance and sustain agricultural productivity, agroforestry can also alleviate some of the pressure for the conversion of forest land into farmland. Moreover, the integration of farmers into forest management schemes through the use of "compromise" land-use systems based on agroforestry may be one of the few realistic ways of sustaining forestry production on agriculturally pressured forest land. A look at some problems and also at opportunities related to the issues of land use and tenure will show that agroforestry has much to contribute to their improvement.

AGROFORESTRY : DEVELOPMENT PATHWAYS

Different agroforestry options and development pathways open up from different stages in the intensification sequence.

Integral Taungya

In the classification of swidden systems, a distinction is made between "integral" and "partial" swidden. Partial swidden reflects "predominantly only the economic interests of its participants", whereas integral systems "stem from a more traditional, year-round, community wide, largely self-contained, and ritually sanctioned way of life".

By analogy with integral shifting cultivation, the proposed concept of "integral *taungya*" is meant to invoke the idea of a land-use practice that offers a more complete and culturally integrated approach to rural development: not merely the temporary use of a piece of land and a poverty-level wage, but a chance to participate equitably in a diversified and sustainable agroforestry economy.

The social aims of the proposed approach are high, and they are nowhere yet fully realised in practice although perhaps the "forest village" schemes in Thailand come closest to the ideal. In some variants of this approach, participation in forestry is made more attractive to traditional shifting cultivators, not only by encouraging them to grow long-term perennial cash crops by widening the between-row spacing of the commercial forest species, but also by allocating permanent agricultural plots for them to use as they see fit.

In addition, they are paid decent wages for a variety of work opportunities and are provided with a range of extension and community development inputs such as housing assistance, clinics, schools, and places of worship. Far from being an exploitative practice, this Thai variant of the *taungya* system promises to become a model example of "integral *taungya*", although nowhere is it adequately documented in the literature. There is always the danger, of course, that the ideals of this approach could be subverted and that the banner of "integral *taungya*" could be used in some places as a cover for politically motivated strategies of ethnic containment and confinement in villages.

Enriched Fallows

There are two variants of this approach: *economically enriched fallows*, which increase the economic utility of the fallow vegetation by enrichment with trees valued for cash or subsistence purposes; and *biologically enriched fallows*, which enhance and accelerate the vegetative regeneration of soil fertility and control of weeds. These become attractive to farmers somewhere around stages 1-2 and 2-3 respectively in the intensification sequence. Long fallow forest shifting cultivators are not likely to be interested in techniques concerning soil fertility and weed control since these are not yet much of a problem, although they may well respond to opportunities for economic benefits from improved fallows.

From the viewpoint of the shifting cultivator, the forest phase of the *taungya* cycle is equivalent to the fallow phase of the swidden cycle. To marry the two types of production successfully the phase lengths must match. Hence, the entry point of primary feasibility for integral *taungya* would seem to be in stage 1, when fallow lengths are of the same order of magnitude as the growing period for commercial forest trees. However, it could possibly be introduced at stage 2, providing that short-rotation forest species are selected, or that agricultural practice can be intensified to accommodate the reduction in available fallow land that would accompany the planting of longer-duration forest trees.

If fruit-trees are planted instead of conventional forest trees, very long "fallow" durations may be required to sustain shifting cultivation, since farmers will be reluctant to cut down the fruit-trees in the years of their heaviest bearing (which may extend from ten to 100 years for some trees). Indeed, the planting of fruit- or other trees - economically valuable, long-lasting and continuously productive - may be the shortest route to permanent removal of the land from the fallow cycle. This is not necessarily a bad thing if the resulting land-use mix, including biologically enriched short fallow or other means of meeting the production requirement of preferred field crops, is significantly more rewarding than the continuation of conventional swidden practice.

The validity of the economic enrichment approach has been amply proved by such indigenous examples as the fallow enrichment planting of rattan by the Luangan Dayaks of Borneo, of cedar and bamboo by the Lingnan Yao of China, of *Casuarina* by the Siane of Papua New Guinea, of gum arabic in the Sahel and of multipurpose fallow wood-lots - whose species diversity exceeds that of the natural forest - by the Ifugao terrace-builders of the Philippines.

The most obvious implication of these practices for tenure is the increased premium their adoption would place on rights of exclusive harvest and reuse by the management unit that undertakes investment of labour for fallow improvement. Without secure rights of this type, it is unlikely that the benefits would be worth the effort. Yet the potential benefits of the approach would seem to justify efforts to bring about supportive tenure changes, always with the qualification that the system could be abused by individuals as a strategy for grabbing land.

Alley-Cropping and other Intercropping Systems

If the preceding discussion stretches the conventional concept of "fallow" beyond its normal usage, the extension of the approach into the "continuous fallow" processes of alley cropping goes even further. It represents a thoroughly functional reinterpretation of the concept for tropical conditions. The beneficial effects of soil-restoring trees on agricultural crops can be effected by associating the two components in *time*, as in the sequential practice of fallow rotation; or in *space*, through simultaneous association of trees and field crops. Alley cropping may be defined as a "zonal" approach to agroforestry, in which field crops are planted in the alleys between hedgerows of nutrient-cycling trees or shrubs.

These are kept pruned throughout the cropping season to control shading and below-ground competition and to provide green manure and mulch material for the benefit of the associated crops. Fodder and fuelwood might be taken as by-products of the system, but the basic aim is to fulfil a "service function" within the arable farming system. The term "alley cropping" was coined by researchers at the International Institute of Tropical Agriculture in Ibadan, but the technique itself appears to have originated in indigenous practice some five decades earlier on the island of Timor under the direction of the Raja of Amarasi.

Figure 1. Alley-cropping

The economic benefits of various experimental alley-cropping systems have been examined by Raintree and Turay, Verimumbe, Knipscheer and Enabor and Ngambeki and Wilson. It has been described in a recent FAO publication as "possibly the most versatile, effective, and widely adoptable of recent innovations in conservation farming". More intimate "mixed" arrangements of trees and crops are also found in traditional practice, the outstanding example being the association of *Acacia albida* with dryland grain crops in the Sahel, where the yields of crops grown in proximity to trees are typically double those of crops grown in the open.

Nitrogen fixation, microclimatic benefits and the peculiar "reverse phenology" of farmers the tree - which leafs out in the dry season and drops its leaves at the beginning of the rainy season, thus nicely accommodating itself to the requirements of crop cultivation - account for part of the yield increase.

Another factor, which has important implications for land tenure, is the concentration of manure in the vicinity of the trees as a result of the livestock that gather under their shade in the dry season to consume the nutritious pods produced at that time. Presumably, the benefits to the arable crops would be reduced if traditional grazing rights were to be restricted.

Security of Tenure

Since the deliberate establishment of new alley-cropping or other, functionally similar, intercropping systems may represent a substantial investment of labour and other resources, security of tenure becomes an important precondition. This does not necessarily mean that the benefits of chosen multipurpose trees selected could not conceivably be shared by claimants with different land and tree utilisation rights. But it is obvious that incentives for adoption of these more-or-less permanent improvements would be more convincing if exclusive-use rights were granted - although it might be advantageous to allow an exception for controlled grazing by livestock during the dry season. In so far as the planting of trees establishes a legal claim to the land on which they are planted, the management unit in question will, in most cases, also have to be the land-holding unit.

For these reasons, and others associated with the relatively higher labour requirements of the practice as compared to planted fallows, intensive alley-cropping systems are not likely to become very attractive to farmers until the short fallow or permanent cultivation stages of the intensification sequence are reached. Then, ecological demands and tenure adjustments make it necessary and possible. Assuming once again that the system is not merely used as a way of grabbing land, supportive tenure adjustments would seem legitimate. And alley cropping itself may provide a technical means of making tenure reforms work. One way of affecting a smooth adjustment of agro-ecological and tenure factors associated with alley cropping would be to phase the system in, based on the concept of an "optimal pathway of intensification".

Starting with a fallow enrichment approach at stage 2, tree species with both economic and biological fallow improving properties could be introduced. By planting the selected trees in hedgerows at appropriate spacing between rows the way would be clear for an intensification of the fallow practice into semi-permanent or permanent alley cropping at stages 3 and 4. Finally, at stage 5, as population pressure intensifies, the children or grandchildren of the original shifting cultivators could install "green manure factories" and maintain a variety of economically valuable upperstorey trees. In this last phase of intensification the system might come to resemble the architectural complexity and economic efficiency of the multistorey home garden so often found in densely settled areas of the tropics.

If the motivation exists, there is no inherent reason why the scheme of intensification envisaged here could not be run "ahead of itself" to generate higher incomes for industrious rural families well in advance of necessity. Of course, this may result in income disparities, but this would be true of virtually any innovation that enhances productivity. The creation of income disparity should not be used as an argument to crush innovation although projects could take measures to ensure equal access to opportunities for adopting innovative systems. Such measures may necessarily involve land reform.

Tree Crop Alternatives

Even with the improvements described above, there is a limit to what can reasonably be expected from the intensification of tropical field-crop systems as an enhancement to the use of fertilisers and other modern agricultural inputs. Several writers have argued the superiority of tree-based cropping systems over arable cropping systems for difficult farming situations, but nowhere is the argument more compelling than in humid tropical environments where the temperate cropping system model has established an unjustified hegemony over the imagination of agriculturists and land-use planners.

The really critical problems to be overcome, then, if high productivity resources exploitation systems are to be developed for these life zones, would seem to be those of achieving an efficient economic utilisation of the diversity of massive, fast-growing, perennial plant species and vegetational types which these environments are themselves efficient at producing. The viability of production systems in which trees and other perennials, rather than annual field crops, are the main focus of the land-use economy is indicated by the widespread practice of "home gardens". In his study of Ibo farming practices in Nigeria, where population densities may be as high as 1000/km^2.

Lagemann found a strong correlation between population density and the importance of multistorey compound gardens. This is largely explained by the fact, that the output from compound gardens is five to ten times greater in monetary terms than from the outfield plots.

The returns to labour are four to eight times higher. In Java, where various tree-garden types exist in a mosaic with wet rice paddies and rainfed arable crops, home gardens may provide more than 20 percent of household income and 40 percent of household caloric requirements.

Highly integrated tree-based economies have been reported from Indonesia, involving the near total exploitation of the lontar palm (*Borassus sudaicus* Beccari) on the islands of Roti and Savu and the domestication of *Shorea javanica* in southern Sumatra, a species which silviculturists have regarded as too difficult for plantation purposes. The role of trees in these economies is exceeded, perhaps, only by the tree-based systems of the Pacific atolls and low islands, where even drinking-water may be provided by a tree - for example, the coconut palm. In this connection, the under-exploited potential of the palms is so striking as to deserve special mention.

Horticultural tree crops for cash purposes are extensively planted by small landholders as an outgrowth of shifting cultivation in many parts of the world, notably the oil palm, cacao, coffee and cola nut plantations of West Africa, covering as much as 67 percent of the land in southern Nigeria; and the coconut, rubber, oil-palm, cacao and coffee plantations of smallholders in Southeast Asia.

The transition to tree-crop-based systems is not equally feasible from all stages in the main sequence of intensification in tropical land use. There are few ecological constraints in stages 1 and 2 to the planting of extensive areas to tree crops, although the economic incentives will generally have to be rather attractive since leisure time is likely to be highly valued in integral swidden societies at this stage of development.

Nevertheless, as Dove has pointed out, extensive cash cropping of trees is a common feature of many relatively long-fallow swidden systems. At stages 3 to 5, however, the transition to tree crops is less easily achieved, because of the commitment of land to other uses and the relatively long lag between planting and first harvest. Here *taungya* practices can ease the burden of the establishment phase by providing early returns of interplanted field crops.

Since the planting of trees as cash crops will often take land out of food-crop production, extensive plantings after stage 1 must usually be accompanied by some form of field-crop intensification.

Needless to say, the investments incurred in tree planting require secure and fairly exclusive tenure by the planters over both the trees and the land on which they are planted, although it is again conceivable, as often happens in Africa, that various usufruct and harvesting rights could be subdivided among different claimants. For example, grazing rights could be exercised by livestock keepers, with or without payment to the landholder, once the critical tree-establishment phase is passed.

Interstitial Tree Planting

Tree planting need not always compete with field crops for land. The planting of fertility-enhancing trees is one case where crops may actually benefit from association with trees.

Examples of such "complementary" economic relationships abound in practice, but they remain underdocumented and little understood, and planners continue to speak of the assumed competition between trees and food crops as if it were, alas, an inescapable fact of life.

Examples of "supplementary" economic relationships, in which trees and other crops have negligible or neutral interaction, can also be cited. Even when there is a "competitive" relationship between trees and other crops, a certain amount of controlled mixing may be justified in terms of the net economic yield of land-use systems oriented toward the diversification of production.

The planting of trees in "interstitial" locations within farms, along farm boundaries and internal borders, or along roadsides, watercourses, and on wasted or underutilised lands in the general landscape, offers a special opportunity for supplementary production. Plantings at these locations, almost regardless of the biological competitiveness of the trees, may be undertaken with little or no opportunity cost.

A recent aerial photo analysis of a watershed in a fairly densely settled farming community in the subhumid midlands of Kenya indicated that, if existing linear features of the landscape—pathways, watercourses, farm boundaries and internal borders—were fully utilised for planting of appropriate trees and shrubs, some 50 percent of the fuel wood and 40 percent of the fodder requirements of the households in the area could be supplied by these hedgerows, with very little competition with existing agricultural land uses.

Can it be assumed that the tenure issues arising in connection with interstitial plantings will be as benign or readily soluble as the technological problems? Possibly not, since, for example, boundaries per se might be a source of dispute, and trees planted on degraded or "underutilised" lands in the general landscape might arouse the concerns of those currently enjoying gathering or grazing rights there. Perhaps some kind of common property approach to multiple use of these lands would be feasible. To satisfy such contrasting interests the use of appropriately selected multipurpose trees might be advisable.

One social organisational approach that might be worth exploring is the partitioning of planting responsibilities and exclusive harvest rights among individual members of an interhousehold working group, organised on a neighbourhood basis and operating at a larger-than-farm scale.

As regards boundary disputes *per se*, it is important to remember that trees often take on legal meaning as boundary markers. In the Kakamega District in western Kenya, for example, it is irrelevant where a barbed wire fence is located since the boundaries of a farm are always judged by the location of the obligatory *Euphorbia tirucali* hedge. Even where boundaries are well established, boundary planting of trees may lead to problems with the neighbours.

There are, perhaps, two approaches to the resolution or avoidance of such conflicts. One is to plant only valuable fruit- or fodder trees and allow the neighbours to take the share of the produce that extends into or falls on their property. The other approach is to choose trees that are as neutral as possible, offering little shade or other competition to whatever might be on the other side of the boundary.

The disadvantage of the second approach is that it may eventually mean that only useless trees can be planted, as in the case of the *Euphorbia tirucali* hedges. Although they can be used for emergency fuel, there are many superior fuelwood species to choose from. Nevertheless, it seems to be the very neutrality, indeed the comparative uselessness of this tree, which gives it its unique legal significance as a boundary marker.

Where land-use patterns and tenure rules are undergoing changes in adapting to population pressures or other factors, the boundary-marking role of trees can have either positive or negative social effects, depending on who is planting them to establish what kind of claims to land, and whether or not such claims are considered legitimate. Trees can be used to consolidate tenure aspects of ecologically necessary and beneficial changes, but again they may also be used for out-and-out grabbing of land.

The planting of trees at interstitial locations within farms might seem to be a wholly positive development, since this is one relatively painless way of increasing the supply of tree products for household consumption, sale or savings on land that is under the direct control of the household. Unfortunately, this is where many of the gender-related tenure problems arise. Everything depends on what kind of trees are planted and where. In the Central Highlands of Kenya, there are men's trees and women's trees.

In Kakamega all trees are owned by men, and there are strong cultural prohibitions against the planting and felling of trees by women. It is said that if a woman plants a tree she will become barren and her husband will die. Women get around these restrictions, of course, by various ruses, and the wisdom of these cultural rules is beginning to be openly questioned.

AGROPASTORAL INTERACTIONS

The ecological problems posed by the dry regions of the tropics make the integration of trees into land-use systems especially imperative and, at the same time, rather more difficult to achieve than in the wetter zones. Two factors are primarily responsible for the difficulties: the first is aridity itself, which increases the risks and the costs associated with successful tree establishment; the second is browsing damage by livestock.

The investment requirements associated with the former imply the need for security of tenure over the trees, and the latter implies land-use conflict over customary grazing rights. Providing that ways can be found to solve tenure problems, the integration of trees into land-use systems in the dry zones offers a number of possibilities for improving the linkage between agricultural and pastoral elements of the economies of these areas, both within and between management units. Moreover, the trees themselves, properly selected and managed for multiple benefits, may provide the means of resolving or at least mitigating some of the most prominent tenure conflicts.

Mixed Farming Systems

The expansion of farming into fragile dryland environments poses special problems for crop husbandry, including conservation and efficient use of limited soil moisture, maintenance of soil structure and fertility, control of wind and water erosion, and provision of feed for draught animals. One of the keys to coping successfully with limited soil moisture is the maintenance of adequate soil organic matter and nitrogen.

In this respect, a recent analysis of the role of nitrogen in plant water use efficiency by Felker *et al.*, has suggested that nitrogen may be more limiting than rainfall in many arid areas. But conventional methods of green manuring and mulch farming have generally been limited by the difficulty in growing enough herbaceous biomass to return nitrogen to the soil, and by the competition with food crops for water, land and labour. Multipurpose trees, particularly the nitrogen-fixing species so well adapted to dryland conditions, offer several advantages over herbaceous sources of organic matter, nitrogen and fodder.

They are generally more drought-tolerant than herbaceous plants. They have a higher feed value during the dry season - particularly the pod-producing species - and are thus better able to ensure the strength of draught animals at the beginning of the rains. They can be grown at interstitial locations on the farm or in association with crops without replacing them. If appropriately selected and arranged with respect to crops, they can offer microclimatic benefits such as windbreak effects and reduced evaporation. They can produce food, fuelwood, building materials and other directly useful by-products while performing their service roles on the farm.

Lastly, as a form of standing capital, they can serve as a source of convertible savings for emergency needs, including what is known as "famine foods". In this latter role, trees can be a partial substitute for livestock, whose main role in Africa is "savings on the hoof". But they can also strengthen the role of livestock on the farm by enhancing the fodder-manure linkage and, if used as living fences, by providing an affordable means of reducing crop damage by uncontrolled grazing.

These benefits can also be applied to livestock owned by others. In the case of interactions between farmers and pastoralists, the growing of additional, high-quality dry-season fodder on the farm and the use of living fences to control livestock access could potentially go a long way toward relieving the main sources of tenure conflict between the two land-use systems. The main constraint to the realisation of these benefits is, of course, the need to restrict livestock access, often by social means, during the establishment phase of the trees. Thus, although we may be able to envisage a partial "technological fix" for certain agropastoral tenure problems, their ultimate solution depends upon social change.

Pastoral Systems

While trees may constitute an effective element in reducing disagreement between pastoralists and farmers, pastoralists are not well advised to wait for farmers to come to their rescue by planting trees. Although it will usually require a much greater accomplishment of "social engineering" to promote effective tree planting by pastoralists, there are a number of pastoral situations in which the need for trees would seem to loom so large as to justify cautious optimism for the success of well-planned projects.

One of these situations is overgrazing around dry-season water sources. The Ferlo Project in Senegal and the Land Management Near Wells Project in the Niger have been exploring various technical and social models for mitigating the ecological effects of herd concentrations around boreholes. Another common situation is overgrazing around pastoral home areas in addition to the problems associated with the provision of feed to young and sick animals that are kept inside.

The Masai *olopololi* is a dry-season grazing reserve maintained jointly by several households for such animals in the vicinity of their residential *bomas*. Since they are protected from grazing during the rainy season, the *bomas* offer scope for the planting of fodder trees for supplemental feed. Some Masai households in the "group ranches" are beginning to experiment with crop production in small, thorn-fenced gardens near the *bomas*. These would also be natural sites for the growing of supplemental fodder trees to meet dry-season feed requirements. These are the kinds of simple agroforestry suggestions that might be worth exploring with pastoralists.

In so far as they are oriented toward the development of small areas under the control of one or a small group of households, they would not seem to pose serious tenure difficulties. The challenge of more general range improvements for the benefit of the mature herd animals presents much more serious problems since it would involve common grazing lands. However, to cite the Masai example again, such pod-producing legumes as *Acacia tortilis* occur naturally on the open range in Masai territory and in some cases their exploitation is controlled by the local group - nowadays the group ranches. According to the accounts given by Masai informants, the "courtesy" rights accorded by one group ranch to another to utilise the nutritious *Acacia tortilis* pods underwent an interesting change during the recent drought.

Normally the neighbours were allowed to shake the trees, causing more pods to fall to the ground for their animals, but as the drought became more serious, the right to the pods was restricted to those that fell to the ground naturally. If, in fact, the Masai are exercising this kind of regulatory control over the use of tree pods, might they not then be encouraged to undertake steps toward the domestication and artificial propagation of this valued range resource?

Pod Propagation

But who would undertake such tree planting for the general benefit? As W.R. Bentley has observed, the "tragedy of the commons" in India is not so much a problem of overexploitation as one of underinvestment. Posing the problem in this way suggests that one approach to encouraging greater investment in the commons might be to identify cheaper methods of investment. It is a fact that such valuable rangeland trees as *Prosopis* spp., *Acacia albida* and other pod-producing legumes can be and often are propagated by livestock.

The predominant tree in the arid rangelands of Rajasthan in India is *Prosopis cineraria*, an introduced exotic which spread naturally, quite without the benefit of tree-planting projects. What is to prevent a rangeland management project from experimentally supplying a quantity of such pods to pastoralists to feed their herds? No doubt the establishment percentage would be low, but so would the herder's investment; the result, which would be known long after the project had come and gone, might be quite significant. It is so far unknown whether this would work because no project seems to have tried it. The point is that there might be some very simple technological fixes to some of the problems of the commons.

At the other end of the scale is quite another kind of intervention in pastoral rangelands. Interest in dryland biomass energy plantations has quickened in recent years. From the pastoralist's point of view, the problem with such schemes is that they tend to represent yet another assault on traditional tenure rights, but is it really inconceivable that pastoral populations could participate in integrated plantation schemes?

The problem of identifying a basis of shared interest would seem, from an agroforestry point of view, to reduce it to the problem of identifying the best-suited multipurpose trees. Why not choose an energy tree from among several outstanding desert biomass producers that also happen to produce copious quantities of high-quality dry-season pods? Again, *Prosopis* would seem to exemplify the appropriate ideotype.

The reason for choosing a pod or other fruit-producing fodder species is that the utilisation of this by-product need not reduce the eventual woody biomass harvest of the plantation; it could make a big difference in the carrying capacity of the range in dry season; and such a choice might make a great difference to the survival rate of the trees in pastoral areas. In this connection it is interesting to note that energy experts are now coming to the conclusion that if biomass energy sources are to realise their full economic potential, they should be approached in terms of by-product and co-conversion schemes.

REFERENCES

Denholm, J. 1991. *Agroforestry in the mountain areas of the Hindu Kush-Himalayan region.* ICIMOD Occasional Paper No. 17. International Centre for Integrated Mountain Development, Kathmandu.

Franzel, S. Coe, R. Cooper, P., Place, F. and Scherr, S.J. 2001. *Assessing the adoption potential of agroforestry practices in subsaharan Africa.* Agricultural Systems (in press).

Kane, B.T., Wilson, G.F. & Lawson, T.L. 1984. *Alley cropping: a stable alternative to shifting tiny cultivation.* Ibadan, Nigeria, IITA. 22 pp.

Mann, H.S. & Saxena, S.K., eds. 1980. *Khejri* (Prosopis cineraria) *in the Indian desert its role in agroforestry.* CAZRI Monograph No. 11. Jodhpur, India, CAZRI. 93 pp.

Olofson, H. 1983, Indigenous *Agroforestry Systems. Philippine Quarterly of Culture and Society,* II: 149-174.

Schirmer, A. 1983, *The role of agroforestry in the Pacific. Suva, Fiji, Univ. South Pacific.*

Van Crowder, L and Anderson J. 1996. *Integrating Agricultural Research, Education, and Extension in Developing Countries*

11

Policy Issues in Forest Plantation Development

This chapter provides an outline of some of the main issues and variables that will determine whether, and where, forest plantation investment is likely to occur in the future. It presents a broad comparative analysis of different economic mechanisms, incentive structures and policies relating to forest plantations around the world, along with a discussion of the main reasons for governments to encourage forest plantation development. The analysis also explores some of the economic challenges to forest plantation development such as risk, viable size and location, and cost and pricing issues.

Future patterns of private sector forest plantation investment (i.e. excluding direct state planting) will largely depend on comparative advantage. However, when considering comparative advantage, two different types of comparison must generally be made:

- a comparison between countries; and
- a comparison between forest plantation investment and other forms of investment in each country.

Considerations such as: the relative growth rates of forest plantations; harvesting and product conversion efficiencies; economies of scale; and relative costs and prices, are most relevant to the first type of comparison. Broader economic considerations, such as: the overall return on the forest plantation investment; investment risk; and liquidity, may be more relevant to the second consideration.

Incentive structures will certainly play a significant role in determining where comparative advantage lies. Thus, the analysis also examines countries where government intervention (either in the form of direct public-sector planting or incentives for private-sector planting) is responsible for much of the increase in forest plantation areas. It also discusses the potential role for forest plantations to meet countries obligations to reduce net carbon emissions under the Kyoto Protocol and the effect that this might have on the outlook.

MOTIVATIONS OF PLANTING TREES

Before getting into detail, it is first useful to note that there is a wide range of reasons for planting trees and that the public and private sectors often have different reasons for doing so. In general, the primary motivation for the private sector to plant trees is to generate financial benefits from their investment. The benefits from planting trees may be direct, such as income from roundwood sales, or they may be more indirect. Examples of the latter include:

- the benefit of having a secure wood supply for a forest processing facility;
- increased security of land tenure (where investing in forest land establishes some sort of long-term right to the use of land); or
- the financial security of owning a tangible asset such as a forest plantation.

In certain instances, government incentives provide a significant proportion of the financial benefits from forest plantation investment.

It is also worth noting that the reasons for planting trees can differ between different types of private sector investor. For example, many individual private investors see forest plantations as sort of savings account. They invest directly in forest plantation establishment or in young forest plantations, which they then manage at a low level of intensity and harvest as and when they require a little extra income. Forest plantations managed under such circumstances are often not managed according to strict economic criteria and may be utilised for non-market benefits. This often results in relatively low harvesting intensities and the use of long rotation lengths.

On the other hand, corporate investors in forest plantations tend to seek higher, more immediate and more stable financial returns from their investments. They are more likely to manage their forest plantations intensively and use economic criteria to determine optimal rotation lengths and make decisions about silviculture. They also often develop a portfolio of forest plantations of different ages that will generate a relatively stable supply of roundwood and, consequently, income. They may do this through planting or the acquisition of already established forest plantations.

Governments support forest plantation development either directly, through planting by state forestry companies, or indirectly, by providing, grants, subsidies or other incentives for the private sector to plant trees. Their motivations for supporting forest plantation development tend to be different to those of the private sector in that they often incorporate non-financial considerations into their decisions.

In terms of the economic benefits from forest plantation development, governments may support development for purely financial reasons, but they will often provide support for broader economic benefits. For example, government support for forest plantation development can be used to promote the development of a forest industry, either at the national level or in particular regions. It can also be used to create a critical mass of forest plantations at which the private sector starts to take an interest in making such investments themselves.

Governments also support forest plantation development to generate social and environmental benefits. Indeed, such objectives are often stated as the main reason for government support in this area. The creation of employment opportunities, particularly in rural areas, is one important social reason that is often given to justify government support for forest plantation development. In Europe, forest plantation development is encouraged to remove agricultural land from production (an objective that can be supported on economic, social and environmental grounds). Governments may also support forest plantation development for the production of non-market benefits, such as: watershed protection; improved visual appearance of the landscape; carbon sequestration; the provision of outdoor recreation opportunities; and land rehabilitation or reclamation. Where these objectives are important, the way in which forest plantations are established and managed is likely to be very different to what would happen if they were privately owned.

ADVANTAGES OF FOREST PLANTATION

A number of economic theories have been developed that explain why some countries perform better than others in different sectors of the economy. These theories suggest that, in the long-run, competitive and comparative advantage will determine how successful a country is in a particular sector. It is, therefore, worthwhile examining what these theories suggest about the development of forest plantations around the World.

Natural Advantage

The most basic factor underlying a country's success in a particular sector is natural advantage. This arises where there are elements of a particular country's natural stock of resources, location, or natural productive capacity that make it easier to produce a particular good or service there rather than somewhere else. In many cases, the development of an economic activity is likely to begin with a natural advantage in that activity. In the forestry sector, the most common example of a natural advantage is the presence of a large natural forest resource. Thus, until recently, natural advantage in forestry existed in countries with large land areas, high levels of forest cover and generally low population densities. Most countries fitting this descriptionhave developed significant forest industries based on this natural advantage.

More recently however, industrial roundwood production from natural forests has been constrained by resource depletion and, more generally, by increased regulation.

One consequence of this is that natural advantage is gradually moving toward countries where trees grow the fastest. In other words, natural advantage in the forestry sector is gradually shifting away from countries with the highest levels of forest resources to countries that have the highest forest productivity. In terms of hardwood species, recently developed fast-growing plantations of *Eucalyptus* species in tropical areas demonstrate a considerable natural advantage in pulpwood production. Similarly, in terms of softwood species, natural advantage is gradually shifting towards tropical and subtropical regions where pine is grown (*Pinus caribaea; P. elliottii; P. merkusii; P. oocarpa;* and *P. patula*) and temperate countries suited to the production of *Pinus radiata* (e.g. Chile; Spain; South Africa; New Zealand; and Australia).

Canada is a country with a large natural forest endowment, but with little natural advantage in forest productivity. Consequently, Canadian forestry practices rely heavily on low-cost establishment options, such as natural regeneration or broadcast seeding, which take several years to re-establish forest cover after harvesting. In contrast, in roughly the same time, intensively managed forest plantations in the four Southern plantation producers can produce a final crop for harvesting. This gives these countries a considerable natural advantage in the production of industrial roundwood from forest plantations.Projections of industrial roundwood production suggest that production in Canada will remain roughly constant over the next decade while, over the same period, production in the four Southern plantation producers is expected to increase considerably.

Competitive Advantage

A more complex theory that explains why some countries are better at certain economic activities than others is the theory of competitive advantage. In terms of competitive advantage, natural advantage is only one amongst a number of factors that determine whether a country has an advantage in a sector. Competitive advantage in a particular sector (at a national level) is generally defined as a country's ability to achieve higher rates of growth and profitability and a larger international market share than other countries can in that sector. The "Porter Diamond" displays the main elements of competitive advantage at the national level.

The Porter Diamond suggests that national competitive advantage can be achieved by bringing together the following four key elements:

1. firm strategy, structure and rivalry—strong domestic competition, forcing firms to develop efficient structures and clear strategies for success, is a core component of success;
2. factor endowments—some degree of natural advantage, such as a large natural resource or a skilled labour force;

3. demand conditions—viable markets exist and these are characterised by strong and efficient competition; and
4. related and supporting industries—a strong supporting infrastructure, enabling cost effective delivery to markets.

Porter argues that long-term competitive advantage is dependent on countries' capacities to innovate and upgrade. These capacities, in turn, arise from competition, pressure and challenge. Innovation interacts with the other important conditions for competitiveness to determine overall advantage. The other key conditions are a strongly competitive domestic market, well developed infrastructure and a network of supporting industries and some degree of advantage, or competitiveness, in the factors of production. However, Porter also specifically emphasises the role of government and the role of chance in determining success.

In terms of the forestry sector, a country's competitiveness in terms of producing industrial roundwood from forest plantations, might then be determined by the following:

1. strong and active competition between domestic forest growers, leading to well-developed plantation management techniques and practices and an industry that is attuned to innovation and enterprise (firm strategy, structure and rivalry);
2. natural advantages in land availability and forest plantation yields, combined with a well-trained workforce (factor endowments);
3. a strong domestic processing and manufacturing industry that is familiar with the types of species grown in the country's forest plantations (demand conditions); and
4. efficient infrastructure, strong forest research capability and a policy environment that encourages enterprise.

The countries that will be most successful at producing industrial roundwood from forest plantations are likely to be those that most effectively develop and bring together this range of attributes. By doing so, they will create a competitive advantage over both other countries with significant forest plantation resources and countries with significant forest sectors based on natural forest resources.

The theory developed by Porter also suggests that, in the long-run, international competitiveness cannot be based on exogenous factors such as exchange rate advantages, low interest rates, or low labour costs. Invariably, these advantages will gradually be eroded over time by a country's success. In the forestry sector this latter point is demonstrated by the strong performance of a number of relatively high-cost, developed countries, most notably in North America and Europe. Of particular interest are examples in the wider forestry sector, of where high-cost countries have developed competitive advantage without having any real natural advantages in the forest sector. Some examples include: the furniture industry in Italy; paper industries in the United Kingdom and the Netherlands; and the particleboard industry in Belgium. Conversely, there are a number of examples of developing countries where the cost advantages of significant endowments of high-value natural forests have been exploited and exhausted, without sufficient investment in developing the other attributes required to sustain international competitiveness.

Comparative advantage

Due to the many infrastructural and technical advantages that developed countries already have, such countries also already have significant competitive advantage in the forestry sector. It might be thought therefore, that these countries would develop significant forestry sectors and that developing countries would be dissuaded from investing in forest plantations. This, however, ignores the importance of the theory of comparative advantage in determining production patterns.

This theory, in essence, suggests that it is not absolute advantage that necessarily determines where that good will be produced. Rather the comparative advantage is of crucial importance. In other words, comparative advantage is held by the country that has the lowest opportunity cost of producing the good in question.

FACTORS AFFECTING FOREST PLANTATION COSTS AND REVENUES

A robust analysis of alternative forest plantation investment projects requires an in-depth assessment of the costs and revenues associated with each alternative.

Land Costs

Land costs are fundamentally dependent on location and, consequently, very high variations in land costs are both possible and likely within individual countries. Land costs will, for example, depend on topography and a range of other geographic and economic factors, such as: soil productivity; potential yields of alternative crops; and relative proximity to infrastructure and markets. National estimates of land costs are, therefore, very crude averages of a cost that can vary widely.

Labour Costs

An important component of many forestry activities is the cost of labour. In general, forest management activities such as planting, pruning and thinning are very labour intensive although, in some countries, broadcast seeding is routinely practised and some of these operations may be mechanised. Harvesting operations are generally mechanised, with varying degrees of labour inputs, ranging from relatively high labour and low capital use in many developing countries to high levels of mechanisation in some developed countries. Labour costs vary markedly across countries, but are generally relatively uniform within countries.

There are two central aspects to labour costs that should be considered in an evaluation of forest plantation costs—the wage rate and productivity. Wage rates are generally highest in the most developed countries, where technology and training also tend to make productivity highest. Consequently, high wage rates tend to go hand-in-hand with high capital costs. Internationally, wage rates are distributed along a continuum, whereas technological and systemic advances mean that distinct leaps are possible in productivity.

In terms of productivity, labour inputs per unit of output will vary according to the species planted, terrain, intensity of management and the level of capital utilisation. There are likely to be enormous differences in these factors between countries and even within countries there can be considerable differences in labour productivity in forest plantation operations. Thompson, for example, produced a survey of forest employment in Great Britain and found a marked difference in labour inputs between the publicly-owned Forestry Commission forests and privately-owned forests. These results are summarised in Table 1.

Table 1. Estimates of labour productivity in forestry in Great Britain in 1988-1989

Function, measurement units and type of forest manager	*Amount*
Establishment (work-years per hectare)	0.057
Forestry Commission	0.145
Private owners	
Maintenance (work-years per 1,000 hectares)	1.507
Forestry Commission	4.276
Private owners	
Harvesting (work-years per 1,000 cubic metres)	0.961
Forestry Commission	1.098
Private owners	
Office/Forest employee ratio	0.249
Forestry Commission	0.182
Private owners	

The difficulty of producing broad estimates of labour productivity for a particular country is further demonstrated by Thompson, who noted that:

> "Privately owned woodlands are, on the whole, on better quality land, concentrated in lowland areas and with a higher proportion of broadleaves and, therefore, require a higher level of maintenance."

It could also be speculated that the Forestry Commission uses specialised labour, has greater economies of scale in forest management and is a more capital-intensive forest manager.

Other Silvicultural Costs

Other silvicultural costs include a range of non-labour costs associated with planting, tending, pruning and thinning forest plantations. In general, these costs are relatively modest compared with land, labour and harvesting costs. The cost of planting stock will generally depend on the species and the extent to which the planting stock has been improved in some way. For example, the cost of nursery-grown forest planting stock for small-scale planting in the United States of America tends to range from around US$50 per 1000 seedlings for *Pinus taeda* to US$225 per 1000 seedlings for *Quercus* (oak) species. The cost of planting stock for large-scale planting will generally be lower, because buyers can negotiate better prices with suppliers or, in the case of many government agencies and some of the larger private companies, they may have their own forest nurseries and plant production or improvement facilities.

The cost of fertiliser, herbicide and pesticide applications will depend on the nutrient levels at each planting site and whether there is a need to spray to suppress weeds or control pests and diseases. For example, typical fertiliser costs in New Zealand forest plantations of *Pinus radiata* are estimated to range from zero to US $350 per hectare. Weed and pest control in these plantations is estimated to cost up to $20 per hectare per annum.

In developing countries, other silvicultural costs are generally much lower, but a number of different costs may be incurred. For example, in 1991, the total cost of planting and managing a forest plantation on an 8-year rotation in Sudan was estimated to be US$ 500 per hectare. Water management and irrigation costs accounted for around 47 percent of this total and forest protections accounted for a further 6 percent of costs.

Harvesting and Transportation Costs

Generally speaking, the most expensive operation in a forest plantation is harvesting the final crop and delivering the roundwood to the forest processing plant. Harvesting and transportation costs can vary markedly and will depend upon geographical factors such as location, topography and the length and quality of roads that will be required for the harvesting operation.

Revenues

Naturally enough, appraisals of forest plantation investments are guided not only by costs, but also by expectations of future revenues. A considerable body of literature and analysis has been devoted to predictions of future price levels for forest products and, in general, two main schools of thought have arisen from all of this work. On the one side, some analysts see ever increasing demand for forest products and increasing levels of restrictions being placed on forest harvesting and expect that these will eventually lead to sustained increases in forest product prices in the future. Opposing this, other analysts point to the fact that the scope for increases in forest product prices is limited by competition from non-wood products and that the forestry sector is constantly developing new technologies and management techniques to increase forest productivity and to improve production techniques. Viewed from this perspective, forest product prices are unlikely to increase in the future in general, although prices in some areas and for some types of product may increase if supply can not keep up with demand.

The first view presented above developed throughout the 1980's and early 1990's, when a considerable amount of concern focussed on the likelihood of a forthcoming wood crisis. During this period, wood and fibre supplies from several traditional sources were reduced by harvesting and export restrictions, leading to expectations that this scarcity would continue into the foreseeable future and, consequently, lead to significant and sustainable increases in roundwood prices. For example, in 1992-93, a short-term log price "boom" was fuelled by a coincidence of new harvesting restrictions in the United States of America and Malaysia. At the time, a number of commentators suggested that this price increase represented a structural realignment in timber markets that would be sustained in the future. Largely because of the "windfall" gains made by forest plantation owners selling their roundwood at this time, forest plantation establishment accelerated rapidly in several countries. For example, new planting in New Zealand quadrupled between 1991 and 1994.

More recently, the second school of thought referred to above has gradually gained wider acceptance. This is, in part, because the price increases associated with the 1992-93 log boom have not been sustained. Indeed, the view that forest product supplies will generally increase to meet future demands without significant price increases in the foreseeable future has been reinforced by the results of several recent global supply and demand analyses.

Similarly, analysis of many long-run series of forest product prices would also suggest that, if historical trends are a good indicator of future trends, then the real price of most forest products is unlikely to rise in the foreseeable future.

The real stumpage price of sugi (*Cryptomeria japonica*) has declined over the last 30 years in Japan. Forest planting costs and harvesting wages in Japan have increased markedly, in real terms, over the same period. This has probably reduced the profitability of forestry activities in Japan and, it is believed, has discouraged forest plantation owners from undertaking_a range of forestry operations (such as thinning) in recent years. The long-run historical stumpage and log price series for other species (including: Douglas fir in the United States of America; Radiata pine in New Zealand; and hardwoods grown in the natural forests of Brazil) display trends that are similar to those shown above. To summarise, a large amount of historical evidence would tend to support the view that significant shortages may induce price increases in the short-run or, maybe, in the long-run in specific niche markets. However, for the majority of roundwood production from forest plantations, which is aimed at mainstream bulk commodity markets, assumptions of significant long-run real price increases are likely to be optimistic.

Taxation

In most countries, the taxation of forest plantation investments is quite a complicated subject. Tax regimes differ widely between countries and are often also very different for different types of investor. It would be an enormous task to attempt to categorise the vast variety of alternative taxation regimes and rates of taxation applicable to forest plantation investments in every country in the World.

A very simple way of comparing tax regimes in different countries is to compare the basic income tax rate applied to companies and individuals. Such a comparison is not, however, sufficiently accurate to assess the total level of taxation in different countries. For example, investors in forest plantations may also have to pay a range of other taxes including: additional direct taxation surcharges; indirect forms of taxation (such as sales and excise taxes); land taxes; capital gains taxes; and capital transfer taxes. The ability to offset taxation against other business losses is also likely to differ between countries, as will various other taxation-based incentive structures or concessions that might be available. In terms of the calculation of tax, regulations regarding depreciation or whether taxation is based on cash profits or earnings-accrued can also vary. Finally, the degree of sophistication of tax legislation and the level of monitoring by authorities is also likely to vary between countries. Any or all of these factors can substantially affect the total level of taxation that the forest plantation investor will have to pay.

There are substantial differences between taxation rates in different countries at even this most basic level. For example, taxation rates in the small selection of countries listed here vary from 10 percent to 45 percent. Ignoring the effects of subsidies and incentives, the total amount of tax paid in different countries is likely to range from 0 percent to around 70 percent. The first (and probably most important) point to note about taxation of forest plantation investments, is that taxation is likely to significantly affect the financial return to any forest plantation investment. Although this may seem rather obvious, it is very important to bear this in mind, because the results of most comparative analyses of forest plantation investments are presented without taking into account the effects of taxation. This is particularly the case in comparisons across countries, where it would be complicated to try to introduce the effects of different tax regimes. This common practice (of presenting only pre-tax performance figures) can be highly misleading, both when attempting to compare forest plantation investment opportunities across countries and, in particular, when comparing forest plantation investment opportunities with other investment opportunities within a country. The most important rule to remember when comparing any financial analyses of investment opportunities is that it is post-tax returns that are relevant.

The second point to note is that, on the whole, the tax treatment of investment in forest plantations is generally favourable. For example, one of the major tax advantages that forest plantation investment has over other investments is that taxes are not generally levied each year on the annual growth in the value of standing roundwood. Instead, taxes are generally only paid when the roundwood is harvested and results in revenue for the investor or owner.

An example of the magnitude of this benefit is given in Perley (1992). Perley demonstrates that, for investment in a forest plantation for 30 years in New Zealand, the effect of taxation is only to reduce the real rate of return on the investment from 8.5 percent (pre-tax) to 7.89 percent (post-tax, calculated using a 33 percent tax rate). In contrast, the effect of tax on an annual fixed-interest investment (over the same period) earning a real rate of return of 7.32 percent (pre-tax), is to reduce that rate to 4.1 percent (post-tax). The relative difference between the pre- and post-tax returns on the two investments results from the annual tax charge that has to be paid each year on the fixed-interest investment. Taxes on income from the forest plantation investment only have to be paid when roundwood is harvested. In some countries, investment in forest plantations receives even more favourable tax treatment, when establishment costs can be deducted from other pre-tax income or offset against losses in other parts of the investor's portfolio. Such "tax breaks" have been popular at different times in various countries, although they often lead to unintended consequences. In developing countries they are also generally less effective_than direct cash grants or subsidies, because many small landowners pay little or no tax anyway.

Risk and Viability Issues

Risk can be defined as the probability or likelihood of sustaining loss although, more accurately, it is a measure of volatility around an expected value.Because of the long time period generally involved in investment in forest plantation projects, risk may be of relatively more concern to the investor than would be the case for an alternative investment with a shorter life-span. For example, with a long-run investment in a forest plantation, the effect on profitability of a sustained adverse change in costs or prices will be compounded throughout the rotation. Furthermore, there may be little that the investor can do, in terms of changing management techniques or marketing strategy, to reduce the impact of such an adverse event. On the other hand, forest plantations do offer some benefits from the perspective of risk.

The anticipated profitability of projects over a long time horizon depends crucially on the cost of capital or discount rate and the twin components of risk and return. In forest plantation projects lasting, for example, 30-80 years, a seemingly small change in the assessment of annual risk would, through cumulative or compounding effects, result in significant differences in the expected rate of return on the project. Thus, even at the planning stage, it is extremely important to identify and try to correctly assess the risk associated with any forest plantation project.

Another point worth noting is that cost of capital is not constant across investment projects and different countries or through time, but should change according to the degree of risk encountered. For example, the risk of investing in (i.e. purchasing) a relatively mature forest plantation is likely to be much lower than the risk associated with establishing new forest plantations. Different financial arrangements to support investment in the forest plantation project (e.g. debt or equity financing) also have different levels of risk associated with them. Consequently, any attempt to produce a robust assessment of the comparative profitability of forest plantation projects across countries (i.e. by considering risk) is fraught with difficulty.

Even just identifying the risk variables that should be taken into account is not an easy task. Furthermore, given the quality of much of the existing data and information about forest plantations, it is almost impossible to assigning meaningful and reliable quantitative values to many of the uncertainties associated with a forest plantation project. In reality, most international comparisons of the financial returns from investment in forest plantations ignore the element of risk or rely on very simple indices or "professional judgement" to differentiate between different countries. Certainly, in comparison with the methodologies developed for assessing risk in financial markets, the techniques used to assess forest plantations are relatively rudimentary. To a large extent, this is because there is insufficient data in the public domain to carry-out a meaningful statistical analysis of many of the risks associated with forest plantation projects. However, the risks associated with forest plantation projects can be explained and described in qualitative terms.

REFERENCES

Tewari, D.N. 1995. *Agro forestry for increased, productivity, sustainability and poverty alleviation*. International Book Distributors, Dehradun.

Van Crowder, L and Anderson J. 1996. *Integrating Agricultural Research, Education, and Extension in Developing Countries.*

12

Sustainable Forest Management

The forestry and natural resources community has debated the concept of "sustainability" for decades. A high level of political commitment to advancing "sustainable forest management" was achieved at the United Nations Conference on Environment and Development. Considerable effort has been invested subsequently in intergovernmental processes to define criteria and indicators of sustainable forest management. Beyond this, governments have committed staff, time and resources to develop a programme to advance sustainable forest management. In spite of all these efforts, sustainable forest management remains an elusive goal.

The recognition of the hopes and aspirations of the many stakeholders interested in the future of forests is a positive step forwards. However, this raises the question of: how can this multiplicity of objectives be addressed once it is acknowledged that it is simply not physically possible to manage forests in a way that simultaneously achieves every aspiration? Increasingly complex modelling and valuation methodologies have been developed to quantify the range of diverse products and environmental services that forests can provide, but trade-offs or compromises still have to be made as policy advice moves from theory to practice. The community of forestry professionals and forest stakeholders continues to search for an acceptable method to reconcile different perceptions of the relative importance or value of each forest management objective.

In the allocation of market goods, prices are the main indicator upon which production, consumption, savings and investment decisions are taken. Thus, because so many of the outputs of forest ecosystems do not pass through markets and are public in nature, many argue that market solutions result in far too little investment in the conservation and management of forests. However, to decide on a different course of investment and conservation, the trade-offs implied must be analysed within the framework of a commonly accepted value system. Values determine the "weight" or importance of each one of the objectives in sustainable forest management, but values vary enormously between different people and are seldom expressed in terms that would provide clear operational guidance. To summarise, if there is little agreement about objectives and their relative importance, it is not possible to conclude whether a forest is sustainably managed or not.

Reconciling conflicting objectives for forest management is compounded by uncertainty. It may take fifty years or more for a forest to develop a preferred habitat for certain types of wildlife or to produce timber of a desired size. This is a long time compared to most other crops or manufactured goods and it adds further complexity to the problem of defining sustainable forest management. Technical specialists seldom hold a universal view about the eventual impact of a management practice on the forest ecosystem. Partly, this is because the forest ecosystem is enormously complex but this uncertainty also stems from the fact that it takes years, if not decades, to obtain reliable results from field-tests of interventions.

SPATIAL AND TEMPORAL ISSUES

People have been modifying forests for a very long time. Available anthropological and ecological evidence points to the fact that people have been living in and modifying forest ecosystems to their advantage for centuries.

The spatial and temporal dimensions of sustainable forest management are additional issues that complicate matters. There are wide-ranging discussions, often between stakeholders with similar points of view, about the spatial dimensions of sustainable forest management. Is the appropriate scale for evaluating sustainable forest management a single stand of trees, a watershed, a landscape, a nation, or the world as a whole?

Similarly, the planning period for sustainable forest management is equally ill defined. Should the forest manager be looking for sustainability over 50, 100 or 150 years, or in perpetuity? While the implicit perception is in perpetuity, it is very important to recognise that forest ecosystems are not static. With or without direct human intervention, forests continue to evolve in the face of the forces of natural change - fire, drought, pest and disease. Even without human intervention, the present forest estate is not what would exist in fifty years from now or even in twenty years time in some cases.

Finally, responses to the shifting trends in public opinion about how forests should be managed have to take into account the long production period. In the last twenty years, there have been substantial shifts in opinion about the role of forests in the economy, in the environment and in societies. In principle, alternative options for forest management should be judged against the long-term objectives of sustainable forest management. These objectives should reflect the values of society, but this raises the question of: should these objectives reflect the values of today's society, future generations or a mixture of the two? Often, proponents of sustainable forest management refer to the values, wants and needs of future generations, but is there really any feasible way of estimating these?

EQUITY ISSUES

Finally, the debate on sustainable forest management reflects expressions of concern about equity or the fairness in which the benefits of forests are distributed. While there are many equity aspects to forest management; most reflect the fact that the poorest people in developing countries tend to live in or near forests. The implications for equity of sustainable forest management, however, vary depending on the type of improvement to forest management that is being considered. For example, improving commercial forest management and harvesting would benefit large numbers of poor people (through, for example, reductions in off-site environmental costs) while the costs would be borne by a small number of (usually powerful) stakeholders. However, stopping the widespread clearance of forests by small-scale farmers in order to reduce deforestation could have negative implications for equity. As these examples show, sustainable forest management can have both positive and negative implications for equity.

The relative weakness of forest policy to address equity issues remains a major challenge for the wider implementation of sustainable forest management. The discussion is furthered compounded by strongly held views on the tools needed to resolve these equity concerns: market, policy or institutional reforms. While technical and economic analysis may play some role, the equity of market outcomes is largely determined through political resolution. It remains useful to recall that attitudes and definitions of equity differ, sometimes widely, among cultures and stakeholder interests.

In view of the difficult conceptual issues related to sustainable forest management, it seems likely that it will remain an imprecise concept. Given this, many analysts have suggested that forestry policy should promote forest management decisions that will contribute incrementally to sustainable forest management (whatever that may mean) or that will, at least, avoid forest management practices that are clearly unsustainable. This "continuous improvement process" or "use with minimal damage" can be viewed as a set of "goal posts" that are based on the best information that is currently available and can be used to practically guide forest management decisions. This view is appealing to those, such as forestry policymakers that are more concerned with the practical application of sustainable forest management and less with the theoretical details.

SUSTAINABLE FOREST MANAGEMENT IN MOIST TROPICS

Moist tropical forests generate a significant number of global externalities. However, of all the forest types in the world, moist tropical forests probably also present the greatest technical challenges to implementing sustainable forest management. Many experiments conducted to examine sustainable forest management have taken place in this region. Most, however, have been marred by non-technical obstacles and events that materialised before conclusive empirical evidence of the technical feasibility of sustainable forest management could emerge. Most of the obstacles that have appeared have mainly been related to a number of economic, institutional and social factors. These have appeared not only due to the technical complexity of tropical forest ecosystems, but also due to a range of other factors, including:

- the length of time required for tropical trees to achieve commercial size;
- the use of inappropriate harvesting systems;
- economic pressures to repeatedly log areas in which regenerated trees have not yet matured;
- encroachment pressures from migratory communities that survive by practising slash and burn agriculture; and
- conflicting land use claims, including settlement schemes and infrastructure developments.

Forest management of moist tropical forests for sustainable wood production is technically feasible but that most experiments have not lasted for long enough to eliminate all doubts. In Latin America, various researchers believe that sustainable wood production is technically feasible, at least in some ecosystems. For example, Barros and Uhl contend that sustainable wood production is feasible in the Brazilian Amazon, especially in the floodplains where the diversity of timber species is lower, the volume per hectare of commercial species is relatively high, growth rates are also high and logging does relatively less damage to the remaining vegetation and to the soil. Southgate reports that the forest management system used in the Palcazú project in Peru is probably biologically sound on the basis of the evidence of abundant regeneration after harvesting. However, the project itself could not be sustained because of security concerns and the poor economic results, which were due to a relative abundance of alternative wood supplies in the area. In Africa, plans to practice sustainable forest management have similarly shown promise but have been plagued by political, social and economic turmoil. It is interesting to note that, in the latter two regions, success has not been determined by the forestry sector but by events outside the sector. This is a common occurrence in forestry.

The little evidence that there is generally points to the fact that the forests under examination are either evolving as the researchers expected them to, or that any technical failures could easily be corrected with existing knowledge. In fact, the balance of the technical evidence is quite positive and the numerous studies and projects already completed provide a solid base for technically sound forest management for sustainable wood production in most of the world's moist tropical forests.

In dry tropical forests, the context for sustainable forest management is completely different to that in moist tropical forests. Forests are more highly valued for their non-timber services. There are a few examples of the successful implementation of sustainable forest management (e.g. Nepal and India) and it is generally believed that sustainable forest management is technically possible. However, for sustainable forest management to work in this type of forest it is essential that forest managers secure the participation of rural people, by integrating their activities with the rural economy and with other activities that sustain the livelihoods of local people. In addition, as in the case of other experiences in the tropical region, it is generally too early to judge whether or not the few approaches, which currently appear to be successful, will be sustainable in the long run.

Experience in the boreal forest region is limited by the fact that there have been few attempts to advance from purely timber oriented forest management systems to more ecosystem-orientated sustainable forest management systems. The few attempts that have been made are all fairly recent and are surrounded by long-term uncertainties. However, despite these *caveats*, it is believed that enough technical knowledge is currently available to practice sustainable wood production in the boreal forest, or at the very least, to identify and avoid the most unsustainable practices.

Similar conclusions can be drawn from experience in the temperate forest zone. It is somewhat difficult to generalise from the experience that has been gained in the temperate forest zone because of the many different types of forest ecosystem within this zone. However, it is believed that, in most cases, there is generally enough technical knowledge and experience to manage most of these forests for sustainable wood production and to avoid the most excessive examples of unsustainable practices.

Finally, there is adequate technical knowledge to establish forest plantations that produce a sustainable flow of wood, but there is less certainty about the sustainability of current forest plantation practices in the broader sense (e.g. in terms of water quality and local land rights). There is also insufficient evidence to report on how successful reforestation projects in degraded tropical forests have been or will be in the future.

FOREST RESOURCES AND FOREST PRODUCT MARKETS

The size and composition of forest product markets and forest resources are important factors that influence the ability of countries to implement sustainable forest management. Increasing income and population will continue to result in greater demands being placed on forest resources for the production of industrial roundwood and fuelwood. However, at the same time, these same pressures will also increase the demands placed on forests for the production of non-wood goods and services and for the conversion of forests to other land-uses.

Despite these many and often conflicting pressures on forests however, changes in forest management techniques and forest products processing technologies will continue to present a range of options to resolve this dilemma. These will, in turn, present a range of opportunities and challenges for the improvement of forest management and these are briefly described below.

It is important to realise that the relationship between forest resources and forest products markets is a two-way relationship. For example, changes in resource availability affect delivered wood costs and products prices in the market. On the other hand, changes in markets can lead to new afforestation or harvesting of species that were once considered non-commercial. The following analysis presents the outlook for forest resources first then the outlook for forest products markets, but it should be remembered that each affects the other.

The latest estimate of the global forest area is 3,454 million hectares. Developing countries contain 2,810 million hectares or 81 percent of this area, of which 1,805 million hectares (52 percent of the global total) is in tropical developing countries. Current forest loss is estimated to be around 11 million hectares per year, nearly all of which occurs in the tropics. In contrast, forest areas in developed countries and non-tropical developing countries are generally increasing. If forest loss continues at the same rate, the global forest area would decline to 3,285 million hectares in 2010. As nearly all of this loss would occur in the tropics, the area of tropical forest could be expected to decline by 155 million hectares to a new level of 1,650 million hectares (or 91 percent of its current area).

The factors leading to deforestation are complex, variable and defy generalisation. For example, a recent analysis of the demand for agricultural land indicates that agricultural needs might only account for less than half of the projected decline (about 65 million hectares). Other factors contributing to forest loss include: overgrazing; over-harvesting of industrial roundwood and wood fuel; forest fires, pests and other natural losses; and urban and industrial development. Different factors are relatively more or less important in different areas and there is tremendous variation in they way in which these challenges can be met.

The outlook for forest products markets will affect the scope for implementing improved forest management and will, in turn, be affected by and have an impact on the current and future status of forest resources. In particular, the following four variables will probably have the most influence on future wood supplies:

- the overall area and type of forest resources in the future;
- the proportion of forest resources that will be available for wood production;
- the way in which these areas will be managed for wood production; and
- technological improvements in forest management, afforestation, harvesting and processing.

In addition to these variables, new research, technology, and ever changing political and social forces will continue to redefine the goals of forest management and the methods of meeting these goals. The main challenge for the forestry policymaker will be to adapt to these changes without losing the momentum to improve forest management.

The depletion of the forest resource base suggested above will not be significant enough to have a major impact on the supply of forest products or prices at the global level, although it may have significant local impact in some areas, particularly on the supply of large high quality logs and fuelwood. However, another variable that must be considered is the amount of the forest resource that can be used for wood production in the future.

The amount of the forest resource that can be used for wood production is affected by two broad sets of forces moving in opposite dirctions across the forest landscape. Developments in forest product markets and advances in harvesting and processing technology constantly expand the boundary of the forest area that is economically viable. More remote areas become economically operable or nearby forest species formally considered unuseful suddenly become marketable. On the other hand, government forestry and environmental policies tend to operate in the opposite direction, by placing greater areas of forest into legally protected areas and implementing other regulations that restrict the areas of forest that can be used for wood production.

Currently, slightly more than half of the global forest area is considered as unavailable for wood supply (1,653 million hectares) due to either: legal restrictions on harvesting; low stocking of commercial species; or economic inaccessibility. Legally protected forest areas cover about 300 million hectares (or around 8 percent of the total forest area) and the total area of legally protected areas (of all types) is increasing at a rate of slightly over four percent per year. However, in spite of that and given the magnitude of the forest area that is not currently used for wood production, it seems likely that there is scope to meet demands for industrial roundwood production and for the reservation of forest areas for other non-extractive uses within the foreseeable future.

Although the broad availability of forest resources in the future might not affect future supply and demand by very much, the way in which those resource are managed may have a more pronounced impact on future supply and demand. This is particularly the case in the temperate and boreal forest zones, where the area of forests is stable or increasing but there is currently controversy in some of the world's largest wood producing nations about how these resources should be managed. Two clearly discernible trends are currently identified (which could broadly be called intensification and extensification) and these will each tend to influence roundwood production in different directions.

The main example of the trend towards more intensive forest management is the increasing importance of forest plantations as a source of industrial roundwood supply. The most recent estimate of the global forest plantation area is just under 120 million hectares, or 3.5 percent of the total forest area. Almost half of this area is located in Asia and at least 70 million hectares of this area will be used for industrial roundwood production. Industrial roundwood production from forest plantations currently is estimated to be roughly 370 million cubic metres or about 25 percent of total production. Many of the current forest plantations are quite young and without any expansion in area, it is reasonable to anticipate an increase in output to 560 million cubic metres by 2010 (or roughly 30 percent of projected industrial roundwood production in this year).

Over the longer term, the importance of forest plantations as a component of industrial roundwood supply will depend upon the growth of industrial roundwood production and consumption and the rate at which the forest plantation area expands. It seems likely however, that the share of global industrial roundwood production that will come from forest plantations will remain at least 30 percent and it would not be unrealistic for industrial roundwood production from forest plantations to grow to 50 percent of world production or even more.

Acting in the opposite direction, there is increasing pressure to modify harvesting practices and silvicultural regimes within both natural and planted forests that are used for wood production. Indeed this has been a focus of much of the debate about sustainable forest management. Many of the measures that are likely to be promoted as more sustainable are likely to reduce the volume of wood that can be taken from any particular forest site.

Studies of the impact that such measures might have on total wood production are few, so it is difficult to generalise from their results. However, the few studies that have been completed to date all seem to suggest that harvested volume per hectare with more sustainable forest practices might be reduced by at least 10 percent and some studies suggest reductions of up to 50 percent in the short-term.

While this may appear substantial at first glance, a significant reduction in harvesting volumes per hectare would occur in only a very small proportion of the global forest area. It is important to recall that in any given year or even any decade, only a tiny fraction of the world's forest is entered for purposes of commecial harvest. Even the widespread introduction of less intensive harvesting in tropical natural forests would affect only 20% of global wood supplies. In this case, reduced harvest in the natural forest is likely to be more than offset by the more intensive production expected from tropical forest plantations. Consequently is expected that, on balance, measures to reduce the intensity of forest management and harvesting will not be significant enough to counteract the effect of intensification in other areas. Adoption of measures to improve forest management are not predicted to have a major impact on global forest products markets.

The other main factors that will influence future wood production and consumption and the interaction between forest products markets and the utilisation of the forest resource, are future changes in technology. It is difficult to forecast how efficiency improvements and the introduction of new and better technology might occur in the future. Therefore, the analysis presented below only takes into account two likely developments: the greater use of plantations to supply industrial roundwood; and the greater use of recovered paper as a substitute for wood inputs. There are likely to be, however, a much larger number of technological changes both within the forest harvesting and processing sectors and outside the forestry sector, that will reduce the demand for wood, make wood products more price competitive or cheaper to produce and more environmentally friendly than they already are. Four broad developments are likely to have the most positive impacts on forest management.

Firstly, forest processing technologies are constantly improving to utilise smaller sized trees, currently non-commercial species and recycled material. This presents opportunities to recover more utilisable material each time a forest is harvested and to utilise forest areas that may otherwise be considered as unavailable for wood production due to species composition. It also allows processors to diversify their inputs to include more material from non-forest sources.

Secondly, technology is constantly improving to recover a greater proportion of utilisable product per cubic metre of wood input. Furthermore, improvements in areas such as pulp and paper processing technologies are continuing to increase the efficiency with which other inputs are used in the manufacturing process. These developments continue to reduce waste at source and improve the environmental performance of the forest processing sector as a whole.

The third improvement is in the development of new products to meet a given end-use. The gradual substitution of reconstituted panels (e.g. chipboard) and engineered wood products for traditional sawnwood and plywood is also likely to increase the efficiency of wood use, because the former tend to use less wood input per cubic metre of output.

The final example of technological change is changes outside of the forestry sector that will reduce the overall demand for wood products. Improvements to products such as PVC windows and doors and the expansion of these materials into new product markets will continue to exert downward pressure on the demand for wood. While this may not be a sustainable development in a general sense (wood, after all, is a natural and renewable resource whereas plastic is not), it may reduce wood demand and make the goal of sustainable management much easier to obtain.

Table 1. The top ten industrial roundwood producers in 1996

Country	*Production* (*in million m*^{3})	*Production* (*as % of world total*)
USA	407	27.3
Canada	183	12.3
China	109	7.3
Brazil	85	5.7
Russian Federation	67	4.5
Sweden	53	3.6
Indonesia	47	3.2
Finland	43	2.8
Malaysia	36	2.4
Germany	36	2.4

Industrial roundwood production in 1996 was 1,490 million cubic metres. This is equal to about 0.26 cubic metres per capita or 0.43 cubic metres per hectare of forest. Developing countries produced about 565 million cubic metres of industrial roundwood, or 38 percent of this total. Of this, 300 million cubic metres (20 percent of the global total) was produced in tropical developing countries and the remainder was produced in temperate developing countries.[5] Furthermore, industrial roundwood and forest product production is concentrated in a small number of countries. For example, the ten largest industrial roundwood producers supply 72 percent of the global market.

Wood product consumption in developing countries in 1996 was equal to about 490 million cubic metres of industrial roundwood inputs, or about one-third of total global consumption. The remaining 75 million cubic metres of industrial roundwood produced in developing countries was exported to developed countries in the form of industrial roundwood and wood products.

FAILURES OF MARKETS, POLICIES AND INSTITUTIONS

As the above analysis has shown, the technical knowledge to develop sustainable forest management practices is available and global forest products markets are not a major constraint, indeed they offer a range of opportunities to pursue sustainable forest management goals. However, sustainable forest management is not widely implemented, particularly in tropical forests. This occurs because markets fail to support sustainable forest management practices. Furthermore, market failures are often compounded by policy and institutional failures that make it even more unattractive to manage forests sustainably. New initiatives such as "model forests", forest product certification and the development of criteria and indicators for sustainable forest management are currently working to address some of these failures, but there is still a considerable way to go.

Market Failure

Markets may fail to yield economically efficient investment in natural resource management when:

- there are non-market impacts from production that are not taken into account in private production and consumption decisions (i.e. *externalities* and *public goods*);
- property rights are poorly defined or enforced and investment is discouraged while consumption is often higher than the optimal level (the *common property* problem); or
- there is imperfect competition and producers or consumers, acting out of self-interest, fail to arrive at levels of production that maximise welfare.

Perhaps the most noticeable way in which markets fail in the forestry sector is in their failure to take account of the significant externalities associated with forest management. Forests produce a range of locally consumed goods that are difficult to measure such as food, fuelwood, and construction materials. Forests also produce a variety of services that are hard to market. Some of these are fairly localised: e.g. soil protection, water quality protection, wildlife, recreation, and views. Other services can be demanded both locally and even internationally: e.g. biodiversity conservation and carbon sequestration. These externalities are difficult to manage because they vary physically across ecosystems and because they have different values in alternative economic settings and cultures. Consequently, the best forest management regime for each type of forest can only be determined on the basis of local conditions.

Sustainable forest management requires forest managers to invest in the production of some of these outputs. Some of these investments could be financially viable (e.g. low impact harvesting techniques), but do not occur because of policy failures elsewhere (e.g. where governments set artificially low stumpage prices). Others will result in little or no financial gain to the forest manager (e.g. curtailment of harvesting activities on sensitive forest sites). A number of studies have predicted that the cost of implementing sustainable forest management will be large. However, it should be noted that most of the cost of implementing sustainable forest management is not a financial expenditure, but rather a limitation on revenues through the restriction of some harvesting activities. In other words, sustainable forest management is not unprofitable, but rather less profitable than unregulated harvesting.

The third way in which markets often fail in the forestry sector is where common property problems arise in the management of the forest resource, when multiple parties share the resource. Shared ownership of forests has often led to the resource being degraded (the tragedy of the commons) because people have little incentive to invest in the forest. Typically, this results in over-harvesting and under-management of the resource. One response to this problem is to allocate property rights for all, one or a specific bundle of forest outputs (e.g. water, timber and forage). However, issues of equity complicate these assignments of rights.

Policy Failure

The typical response to market failure is to design and implement government policies to attempt to either regulate production or create markets for some of the externalities. Thus, for example, many countries have a large body of forest legislation identifying areas of forest that can be used for different purposes and describing how they should be managed. Others have subsidies to support desirable interventions and fines or penalties to discourage bad practices. However, other policies within and outside the forestry sector often counteract or diminish the effect of these policies.

One of the major problems in many countries is the incorrect pricing of forest outputs. Such policies, often implemented with the aim of stimulating development of the forestry sector, typically price standing roundwood at a level that is much lower than would be obtained in a competitive market. Pricing policies are often also poorly designed in terms of their structure as well as the level of charges imposed and may not be strictly enforced.

There is considerable evidence of under-pricing and low levels of fee collection around the world. Such policies discourage efficiency in harvesting and processing, discourage the development of alternative wood and fibre sources and lead to inefficient allocation of scarce development resources (land, labour and capital) towards the forestry sector. In other words, they act against the improvement of forest management. An associated problem is bans on the export of raw or semi-processed forest products that depress local prices for felled roundwood and forest products. These have similar impacts to under-pricing of the standing trees.

Most of the subsidy given to the forestry sector (in the form of artificially low prices for forest resources) is captured by forest concessionaires and processors, although some of the benefits from activities such as illegal logging (where no fees are paid) are passed on to consumers. This subsidy supports very inefficient operators at the margin, while the others earn high levels of profit. The forest industry often claims that they can not afford to pay higher stumpage rates or invest in sustainable forest management, but there is plenty of independent evidence to suggest that this is not the case. Consequently, it seems likely that large parts of the forestry sector could afford to invest in better forest practices, even though they are unlikely to be able to pass the cost onto consumers, if only they were compelled to do so.

There are also a number of examples of policies outside the forestry sector having a detrimental impact on the management of the forest estate. Policies that encourage the development of other land-uses such as: mining; agriculture; roads; and urban development, are typical examples. These can often be justified on the grounds of economic efficiency (i.e. they represent a higher value land-use), but sometimes they can not. Broader structural adjustment programmes have also been shown to sometimes have a negative impact on forest management. To some extent, the current debate about sustainable forest management is futile in this respect. Countries will continue to convert forestland to other uses where it is more profitable to do so. Forestry policymakers are largely powerless to stop this. A more constructive approach may be to seek improvement in overall land-use planning and compensation when such changes occur, to support improved forest management in remaining areas.

The other major area of policy failure in the forestry sector is in the legal framework governing land tenure and titling. National laws and regulations sometimes conflict with each other or with local legislation and such laws often fail to take into account traditional or customary laws, which often have more impact on the way in which forest land is managed in remote areas. The uncertainty that this creates further exacerbates the common property problems referred to above. In some cases, laws governing land tenure and titling even encourage deforestation. This occurs in situations where the first steps towards legal property rights over public land forest can be obtained if individuals demonstrate that they have improved or invested in the land in some way. Legal precedence in many countries has shown that one common way to do this is to clear the forest cover and replace it with another crop.

Institutional Failure

Institutional failure occurs where countries have adequate existing forestry policy and legislation in place to support the implementation of various aspects of sustainable forest management, but still nothing occurs on the ground because these policies are not implemented. Indeed, the lack of implementation of existing government forestry policies is often seen as a more important cause of the continuation of poor forestry practices than the quality of the policies and legislation already in place.

In many respects, institutional failure can be seen as an extension of market failure. Governments, often fail to adequately fund forestry administrations to carry out the tasks they are expected to perform and sufficiently train their staff. The common property problems associated with forestry apply to government institutions as much as they do to private individuals. Thus, it is not uncommon to see different government agencies offering rights to perform incompatible activities on the same piece of land. The distribution of benefits amongst local and national authorities also often varies between different sectors. Institutions that try to control operations from the centre and where a large proportion of revenues go to the national government, often fail to successfully implement policies over widely dispersed areas and, unfortunately, forestry administrations often fall into this group.

Another institutional failure often arises because of the imbalance in money, skills and power between the different stakeholders. Forest managers and the processing industry typically have substantially more of all of these assets than government officials and local communities. This sometimes leads to corruption of government officials and local leaders and it often makes it difficult for other forest users to negotiate outcomes that take into consideration their concerns.

REFERENCES

Anderson,J. 1999 "Pluralism and sustainable forestry and rural development". Proceedings of an international workshop 9-12 December . FAO 1997.

Deb Roy, R. & Pathak, P.S. 1983. "Silvipastoral research and development in India". *Indian Rev. Life Sci.*, 3: 247-264.

Dwivedi, A.P. and Sharma, K.K. 1989. *Agro forestry: its potential.* Paper in Seminar on Eucalyptus, FRI, Dehradun.

Eastin, I. 1996. *The role of marketing in promoting a sustainable trade in forest products in tropical regions.* Paper presented at the FAO/UPLB Workshop on Forest Products Marketing, Los Banos, Laguna

13

Agroforestry and Poverty Alleviation

People everywhere are more highly dependent on forests for food security than generally known. The complex linkages between forestry, on the one hand, and food security, on the other, are among the least recognised and appreciated of all food security factors. While food security policies and investments focus primarily on production and trade of major staple grains, the crucial roles of forests in enhancing and sustaining food security are often overlooked or ignored. This subject highlights the intricate linkages between forestry and food security. It also provides an overview of the forestry and agroforestry issues that significantly affect food security in Asia and the Pacific.

People have obtained food products from forests for thousands of years. Fruits, nuts, leaves, flowers, stems, roots, and mushrooms are examples of the many kinds of foods that are gathered from trees, shrubs, and forest ecosystems. Forests are also home for birds, snakes, lizards, insects, and other wildlife commonly consumed by rural people for protein. Trees likewise contribute to nutrition and health by providing wood fuels for cooking and traditional herbal medicine for local use.

In recent years, as more people migrated to urban centers and agriculture generally became more intensive, traditional links between forests and food security have become obscured. It is, nevertheless, important to realise that the most important contributions of forests to food security are not the most obvious or direct. For instance, the most significant food security values of forests and agroforests lie in their rich genetic reserves and in their environmental protection functions. In this sense, modern agriculture and overall food security are perhaps more dependent on forests and trees than ever recognised.

Unfortunately, the capacity of forests to contribute to, and sustain, food security is seriously jeopardised by the rapid pace of deforestation, particularly in the tropical parts of Asia and the Pacific. Immediate and concerted steps are needed to safeguard the remaining forests and restore the degraded ones in order to maximise forestry's roles in contributing to food security.

LINKAGES BETWEEN FORESTRY AND FOOD SECURITY

In some areas, forests and agroforests provide a large proportion of rural households' food needs. These contributions come in two major forms:

(a) the environmental protection role of trees and forests that enhance water and soil conservation to maintain high levels of productivity;

(b) the direct food commodity contributions which can supplement normal farm yields or serve as substitute products in the event of crop failures due to floods, droughts or insect infestation.

Also of major importance are the socioeconomic contributions of forestry to food security.

Watershed Protection Functions Agroforests

Forests are especially important in protecting the natural resource base upon which sustainable agriculture depends. No aspect of this protection is more crucial than the watershed protection functions provided by forests. Without adequate protective forest vegetation that insures sound watershed management, it will be impossible to increase crop yields, or even to sustain current yield levels. Uplands and lowlands are intimately linked in hydrologic terms. Adverse land-use changes and practices in uplands almost inevitably affect the agricultural potential of lowlands for the worse.

Deforestation and poor land-use practices in water catchments greatly accelerate erosion and site degradation, and often lead to reduced storage capacity in reservoirs, lowered irrigation potential, and increased incidence of downstream flooding. Examples from Asia are numerous. High sediment loads of rivers draining the Himalayas, where land use is intense and erosion is high, have caused problems for agriculture in the lowland plains of Pakistan, India, and Bangladesh.

Similarly, intensive land use in watershed areas has accelerated siltation of reservoirs along the Yellow River in China, thus affecting irrigation potential along the river. Responsible forestry practices are key elements of watershed protection and rehabilitation. They include strict protection of all forest vegetation on critical steep slopes, application of low-impact logging practices, afforestation or revegetation for water catchment protection, fire control, and adoption of agroforestry systems to reduce soil erosion. By contributing to sound watershed management, healthy upland forests can help regulate river flows which feed downstream agricultural areas. They may also assist to some extent in reducing the frequency and intensity of floods.

Protection against Water and Wind Erosion

Soil erosion caused by water runoff is a serious problem for agricultural production throughout Asia and the Pacific. It strips away the most fertile top layers of soil, robbing the land of its capacity to produce food and other crops. Forests and trees can provide protection against some types of water-induced soil erosion. It is widely recognised that combining trees with other soil conservation measures can greatly extend the possibilities for sustainable crop production on sloping land.

The incidence of shallow landslides and land slips is greatly influenced by the vegetation growing on slopes. Tree roots can isubstantially increase the stability of slip-prone slopes. Studies in New Zealand have shown that tree roots provide up to 80 percent of the soil shear strength under saturated soil conditions. Trees also protect crops from wind damage. Some of the most dramatic crop yield increases attributable to windbreaks have been reported from China, where *Paulownia* spp. has been planted to protect more than 13 million ha in the windy central Plains. Grain yields increased 60 percent or more in fields protected by windbreaks, and cotton output has tripled.

Support for Farming in Marginal Lands

As population pressures on land have increased in Asia and the Pacific, farmers have been forced to cultivate ever steeper slopes and less fertile sites. Successful farming of such steep, infertile uplands often depends on the use of trees and living hedgerows, planted in combination with food crops. Sloping Agricultural Land Technologies (SALT), developed in the Philippines using hedgerows of nitrogen-fixing trees and shrubs, are now widely applied throughout the region.

In addition to holding the soil in place, the leaves of the hedgerow species are used as green manure to enrich the soil and enhance crop production. Various such agroforestry techniques—purposely combining trees with agricultural crops or animals—are becoming increasingly important in the drive to maintain and expand food production on marginal farm sites throughout the region.

Agroforests as Sources of Animal Feeds

Trees and shrubs indirectly contribute to food production by providing fodder for livestock which supply meat and milk for food, or for livestock used as draft animals in farm production. Fodder-producing tree plantations provide feed either by the "cut and carry" method for stall-fed animals, or through open browsing among free-ranging animals.. As much as 50 percent of all animal fodder in Nepal and parts of India comes from trees.

Agroforests as Direct Sources of Food

People living in and near forests traditionally obtain significant portions of their diets from forests. Forest foods are particularly important in predominantly subsistence economies in remote areas. Dependence on these foods is epitomised by the isolated "hunter-gatherer" communities that still exist in parts of India, Papua New Guinea, Sarawak, and a few other locations in Asia. In the Indian states of Bihar, Orissa, Madhya Pradesh, and Himachal Pradesh, most forest dwellers still depend on the forest for 25 to 50 percent of their annual food requirements. As recently as 1988, people in the wooded areas of northeastern Thailand were getting as much as 60 percent of their food directly from the forests.

In modern times, it is unusual for people to still get that much food directly from forests, but most rural people still depend on forest-based foods to supplement their regular diets. Such forest foods can, therefore, be important elements of sustainable and nutritionally balanced diets. Moreover, they contribute to diet diversity and flavor.

In agricultural areas, the most widespread direct contribution of forestry to food production is through food-producing trees on farm and fallow land and around homes. The extent of this contribution varies widely. At one end of the spectrum are the sophisticated "home gardens" [of which the Javanese home gardens are the best known], which provide more than half of some families' food. At the other extreme, is the single mango tree, or other fruit tree, planted in the family backyard.

Trees and Nutrition

Forest foods rarely supply the main part of a family's diet, but tree and forest products have an important role in ensuring adequate and balanced nutrition for rural people. A recent study of northeast Thailand villages revealed that the nutritional status of preschool children in villages near forests was considerably better than that of children living in villages far from the forests Similar results were found in studies in Bangladesh. One of the most common uses of forest foods, especially fruits and insects, is as snacks. For example, people frequently eat fruit collected from the forest while they are working, herding animals, gathering fuelwood, or tending their fields.

In some areas where crop failures are not uncommon, certain forest perennials are important as "emergency foods," or "famine foods." In India, Malaysia, and Thailand, about 150 wild plant species have been identified as sources of emergency food.

Agroforests as Pools of Genetic Resources

One of the most important connections between forests and long-term agricultural stability is their role as pools of genetic diversity. Modern agriculture continues to depend on periodic infusions of "new" genes from wild relatives to combat outbreaks of disease and susceptibility to insect attack. Major agricultural crops that have "borrowed" significantly from wild relatives include tomatoes, maize, peanuts, potatoes, wheat, rice, and sugarcane. In the late 1980s, for example, researchers in Mexican forests discovered a wild species of maize that is resistant to five major diseases of domestic maize. Many of these wild relatives are found only in forests and the margins around forests. When the forests are destroyed, irreplaceable genetic resources are also lost along with the forest. From the point of view of future agricultural production, the wide variety of species that forests contain—both known and yet to be discovered—may have critical roles to play in providing the genetic variability and biological diversity needed to combat the ever-adapting pests and diseases that prey upon food crops. A major challenge lies in safeguarding and conserving these valuable genetic resources for potential future use in the face of pressing immediate needs by current generations.

Socioeconomic links between Agroforestry and Food Security

Wood is important to food security in many ways. In developing countries of Asia and the Pacific, it is not uncommon for at least 75 percent of the population to depend on fuelwood for cooking. Shortages of fuelwood may mean that food is inadequately cooked, or is only cooked once a day. The excessive time and effort spent by women to gather and transport fuelwood for cooking family meals leaves them very little time to engage in other important home tasks, such as cooking meals, caring for the young children, cleaning house, etc.

In areas where fuelwood is scarce, people may turn to dried animal dung for fuel. The burning of dung deprives farmers of an important fertiliser that would normally be returned to fields to nourish crops. Shortages of fuelwood therefore may not only mean more difficulty in cooking food, but may lead to less food being produced.

Food security implies both physical and economic access to food. Forests in Asia and the Pacific generate huge amounts of income and employment that place people in a better position to purchase rather than produce their own food. Income and employment are generated when people become involved in family- or community-oriented forestry activities such as harvesting, processing and marketing forest products. Common features of such enterprises are that they are small-scale (often employing family labor), flexible, and seasonal (operates only when family labor is not engaged in other farm activities.) A broad estimate suggests that forestry currently provides the equivalent of 60 million work-years worldwide, of which about 80 percent is in developing countries.

Collecting forest products from the natural forests serves as an important income source for many traditional societies. Gathering or harvesting non-wood forest products is especially important. Some of the most extensive examples of forest products gathering come from India, where at least 3 to 4 million people are involved in the commercial woodfuel trade and up to 7.5 million people are involved in the collection of tendu leaves (*Diospyros melanoxylon*) used in making bidi cigarettes.

An important way of increasing the cash-generating potential of forest resources is through the development of processing enterprises. The processing of rattan into furniture, for example, employs at least half a million people in Southeast Asia. Other examples of important export crops arising from the forest include bamboo shoots, honey, spices, gums, and resins.

Cash cropping of trees by rural people is another viable source of income and employment under certain circumstances. In many countries of Asia, farmers are planting large numbers of trees-not for subsistence fuelwood-but for cash income. In some areas, farmers earn far greater returns from growing trees than from traditional agricultural crops. Fruit trees are especially popular in many areas, providing farmers with income from the both the fruits and the wood of trees.

The intricate linkages between forestry/agroforestry, on the one hand, and food security on the other, highlight the sensitivity of forestry issues as they relate to food security in the Asia-Pacific region. Although not always readily apparent, "forestry" issues are often "food security" issues as well, and vice versa.

The Asia-Pacific region is home to more than half the world's population. While population growth has slowed down in recent years in many of the countries, the absolute numbers continue to rise and population densities per area of land remain the highest on Earth. Coupled with the huge numbers of people is the increasing wealth being generated in the vibrant economies of the region. Combined, these two factors are placing unprecedented pressures on the region's limited natural resource base, as people are demanding ever more food and other products.

Not only is more food needed to meet the basic calorific requirements of the expanding population, but increased wealth allows the people to purchase much more meat products than in the past. Meaty diets, of course, require more grain and forage to produce than equivalent vegetarian diets.

The increasing demand for food runs in parallel with the escalating demand for forest products and services, including traditional wood products, water, and electricity from hydro-electric projects. The region's new-found wealth also has spawned new demand for such things as forest-based recreation opportunities, ecotourism, and attractive industrial and residential sites. Local and international demands for biodiversity conservation, protection of endangered species, expansion of protected areas, and carbon sequestration add to the challenges of forest management and limit options for conversion of more forests for agricultural production.

Forest Conversion and Deforestation

Throughout history, agricultural expansion to increase food production has occurred largely at the expense of forests. When more agricultural land was needed, people simply cleared more forested land for crop planting. It is no small coincidence that the area of new land brought under agricultural production each year is almost exactly equal to the area that is deforested each year-approximately 15 million hectares.

For much of the world, however, the limits of the land frontier have been reached. In Asia, especially, there simply is no more frontier in most areas. More than 80 percent of all potential cropland is now under cultivation. In comparison with other regions of the world, Asia and the Pacific has a much higher proportion of its land devoted to agriculture, and, conversely, a much lower proportion under forest cover. In most Asian countries, the areas of easiest access and those with the best cropland potential have long been converted to agricultural production, and any further expansion will necessarily encroach on marginal areas-particularly steep hillsides, infertile sites, and semi-arid regions-that are relatively fragile and have relatively low productive potential.

Conversion of additional marginal areas of these types are likely to have a net negative effect of food production and food security, particularly in the long-term. In most cases the adverse impacts of watershed degradation and soil erosion on downstream agricultural systems will be much greater than the minimal increments of food outputs from newly cleared marginal lands. As described above, the clearing of forested areas carries with it the added risk of further degrading the genetic resource base upon which future agricultural productivity may depend. Despite the critical importance of forests in contributing to food security, the region's forests are being destroyed at an unparalleled pace.

FAO's last Global Forest Resources Assessment estimated that deforestation in tropical Asia and the Pacific now exceeds 3.9 million ha each year. This means 7.4 ha are cleared every minute. Most of the forest losses in the world today occur in the tropics. Of the three tropical regions of the world, Asia and the Pacific has the fastest rate of deforestation, the fastest rate of commercial logging, the highest volumes of fuelwood removed from the forests, and the fastest rates of species extinctions. Countries losing the largest areas of forest include Indonesia, Thailand, Myanmar, Malaysia, Philippines, and India.

The underlying causes of deforestation are rooted in a complex web of social, economic, and institutional problems both inside and outside the forestry sector. These include:

— the combined effects of poverty, inequitable land distribution, insecure land and tree tenure, low agricultural productivity, and rising population pressure;

- increasing demand for tropical timber and agricultural products;
- international debt obligations which push developing-country governments to accelerate forest exploitation to earn needed foreign exchange; and
- misguided policies and incentives which encourage, or fail to restrain, the clearing of forests.

Efforts to reduce or eliminate deforestation will necessarily have to address these problems if they are to succeed.

Trade and Marketing Issues

With increasing domestic and global market demands, trade and marketing in forestry and agroforestry products have penetrated into even remote forest areas. This trend presents both opportunities and problems. It can serve as an impetus for increasing household production, income and food availability. On the other hand, over-exploitation to satisfy growing market demands can lead to irreversible damage inflicted upon the natural resource base.Following UNCED, there has been an upswell of concern for and emphasis on sustainable forest management. National and international criteria and indicators of sustainable management have been or are being developed, as well as mechanisms for certifying that certain products are managed sustainably. Many contentious issues, some polarised in a North-South nature, are also being debated, including: intellectual property rights; trade barriers and tariffs; and trade liberalisation.

Resource Access and Tenure

Of primary importance to food security is access to resources. This includes access to natural forests to collect food and to hunt wild animals. It also includes access to land, secure tenure over food production assets, and access to credit for improving production inputs. Tenure security is particularly important to encourage long-term investments in trees and land improvements to enhance food production.

Food security of local people can be threatened if policies restrict traditional access to badly needed forest foods. On the other hand, unrestricted access cannot be granted to sensitive areas whose destruction would jeopardise the long-term sustainability of downstream agriculture. In this regard, careful land-use planning is needed, backed by practical implementation in the field.

Gender Issues

A key aspect in enhancing food security is the involvement of women. Given the central role that women play in food production and food security, their concerns are vital in guiding forestry and agroforestry development in ways that will effectively address food security. Experience has shown that it is not always necessary to employ women extension agents and establish separate initiatives for women's participation. A better approach may be to simply ensure that women are brought into existing mainstream institutions and planning processes.

Management of Fragile Lands

With ever-increasing numbers of people to be fed in Asia and the Pacific, and increasing land pressures, a reorientation of forestry officials, management priorities, and operating procedures is needed. This will often require compromise on the part of forestry departments away from what is conventionally regarded as "best" from the forestry perspective. It could mean, for example, planting timber trees at wider spacing to allow food crops to be grown for more years before the canopy closes. It could also mean accepting, or even encouraging, the use of non-traditional or fruit-bearing trees instead of timber species for reforestation and watershed protection purposes.

There is an increasing appreciation among resource managers for the traditional knowledge and values that local people have toward forests. This recognition of indigenous forest management knowledge, coupled with frustration over the inability of many governments to solve lingering forest management problems, is leading to the devolution of management responsibility to local communities of local leaders. This trend has major positive implications for food production and food security. Not only are indigenous management systems more apt to be focused on local food production, but they are also more likely to protect the genetic diversity of the forest.

Not surprisingly, such reorientation of management perspectives and priorities is fiercely resisted by some conventionally oriented foresters. Thus, there is a need for strong high-level commitment and extensive retraining of forestry department staff if such management reorientation is to succeed.

Appropriate Technical Solutions

The technical solutions most suitable for enhancing food security through forestry and agroforestry may not always be those envisioned by foresters or extension agents from outside local communities. Projects too often fail because they attempt to push pre-packaged technologies upon local people who may not be willing to accept them. Most people are logical decision makers. If they do not accept suggested technical approaches, it is either because they do not have adequate information, or because something is wrong with the approach. Forestry extensionists and rural development officials must be sensitive to both possibilities and be prepared to adapt technologies to meet local social, cultural, and economic conditions.

REFERENCES

Anderson,J. 1999 "Pluralism and sustainable forestry and rural development". Proceedings of an international workshop 9-12 December FAO 1997.

Shah, P, 1988, Potential of Agroforestry as land use technology for rural poor. Advances in Forest Research in India, 1:27-100 p.

Sharma, K.K, 1996, Agroforestry in farming systems development. The Indian forester, 122 (7): 547-599 p.

Tewari, D.N. 1995. *Agro forestry for increased, productivity, sustainability and poverty alleviation*. International Book Distributors, Dehradun.

14

Industrial Agroforestry

Agroforestry techniques generally operate at one of three levels. *The first level*, and the one currently receiving most attention internationally, is agroforestry practiced by small farmers with the purpose of satisfying family or village-level needs. This may simply refer to one or two single-purpose eucalypts growing in the garden to supply part of the family's fuelwood needs or to half a hectare of *Leucaena*, a few of which are cut whenever a neighbour wants to pay or barter for some firewood. A relatively secure land tenure and tree tenure system is one of the best incentives for this level of agroforestry.

The second level of agroforestry (AF) is that promoted by local governments, usually on a fairly large scale. *Taungya* in its several modified forms is typical, whereby rural people are given incentives to plant trees along with their crops and care for them. At the end of 2-4 years, when the trees are well established and the crops begin to fail because of weed competition, soil nutrient depletion and shade from the trees, people are relocated to a new site and the land is allowed to go into fallow.

Ideally the fallow period lasts until the trees are at the optimum point of maturity. They are then harvested and a new *taungya* rotation begins. The land and trees *may* belong to the people doing the work, but generally this is not so. Stability in tree and land tenure is provided by the local government. This level of AF will fail with insufficient incentives or where the population-land ratio is too low or too high. When it is too low, farmers have enough free land to support themselves without a formal, government-operated system. When it is too high, fallow periods are necessarily too short for the land to recover and for the trees to reach optimum size for commerce or local use.

The third level of agroforestry has received very little attention in the literature, but is a highly promising technique. It consists of large-scale AF efforts, usually by the private sector. It is aimed at maximising outputs (profits) on an industrial scale by the production of wood in combination with crops or livestock on a given piece of land. It may or may not have goals directed toward the benefit of local rural populations. It usually requires a secure land tenure system, the land being either owned outright by a large company or held in the form of a concession or a long-term lease.

It may, however, also succeed under small ownership tenure patterns. This level of agroforestry is applicable to temperate as well as to tropical zones. Indeed, a form of AF has been proposed for the pine forests of the southern United States. This industrial level of AF can make large-scale reforestation much more practical in economic terms.

A wood-producing company can often repay its reforestation costs within 25 years by producing maize, sorghum or other crops among the newly planted tree seedlings rather than suffering interest costs over the entire rotation. Such a scheme can make the use of higher-quality sites, fertilised trees, mechanised tree farming and shorter rotations economically feasible. Establishment costs (especially the associated interest charges) are minimised, site maintenance costs are greatly reduced, and site productivity is maximised.

There are several variations which can be applied. If, locally, there are underemployed farmers or if there is a land shortage, a cooperative venture would serve to meet the needs of rural people in addition to producing forest products on a large scale.

Such a venture would use local labour through a variety of incentives instead of using mechanised, labour-saving techniques. If the local population raises livestock, fodder production may become part of the operation. If land-ownership patterns are typified by small, unproductive farms and the private company interested in wood to harvest has insufficient or no land, it may be feasible for the company and the small landowners to work together to produce the wood, food and fuel needed by all concerned.

Vergara's recent matter on agroforestry concluded that, because industrial firms with forestry projects in developing countries do little to satisfy the needs of rural populations, "forestry practice should be converted from industrial forestry to village or community forestry". There can be no doubt that industrial forestry operations in developing nations have often overlooked the needs of those people living in or near the forest. Lack of involvement of these people has often been directly related to failure of such projects: people who feel left out, uninvolved, forgotten or exploited can hardly be expected to cooperate.

Conversion from industrial forestry to village or community forestry may well be the best option in some situations, but there is another option to consider. Indeed, some large industrial concerns have been modifying their approach to use industrial agroforestry as a technique to work in partnership with rural populations. Other companies have employed industrial agroforestry techniques on their own, presumably in some degree of harmony with local populations but without their direct participation. In a wide variety of socioeconomic situations, industrial AF presents an opportunity for a company to make a solid contribution to the needs of local people, and a profit from industrial production at the same time.

INDUSTRIAL AGROFORESTRY PROJECTS

Crecex, Chile

Agroforestry is not often thought of as being practiced in temperate zones, possibly because temperate crops are grown mostly as monocultures that can often be mechanised. However, there have been—and in some countries, still are—examples of agroforestry practices in temperate climates. Indeed, industrial agroforestry may be as applicable to temperate climates as to tropical ones, or even more so.

A case in point is a system used in Chile in 1981 on the El Tollo Tree Farm of Crecex-Georgia Pacific Ltd, located in the coastal range of southern Chile near Quirihue. The system is a traditional one that probably evolved from slash-and-burn practices.

On the El Tollo Tree Farm a 20-22-year-old stand of Monterey pine (*Pinus radiata*) was harvested over an area of approximately 300 ha. Local farmers who were interested in planting wheat on the cut-over area were recruited. These farmers helped with the slash burning and then selected their individual plots within the 300-ha area. They ploughed the plots with a team of oxen and at the same time burned slash that had been missed in the initial burn. Large pieces of slash were turned perpendicular to the slope to facilitate ploughing: this also helped to reduce soil erosion.

After the ploughing, wheat was sown. Seedlings of Monterey pine from the tree farm's nursery were then planted at 2 x 2 m spacing. Most of the local farmers camped on the site to tend their crop; they also acted as on-site caretakers to deter stray animals from entering the area. The wheat was harvested by hand in the autumn and the tree seedlings were left to continue growing. Although this system was employed for one season only, it could have been continued through a second, and possibly a third season. However, the company discontinued managing the site as well as other planting projects elsewhere because of weakened market conditions.

The system has many advantages. Farmers are comfortable with it as a system. Because of the very high yields that result from planting wheat on land that has been 20 years in fallow, they are eager to participate. To the industrial tree farmer there is not only the advantage of free labour, but also of protection that would probably otherwise be effected only at extra expense. There is also closer relation between the local farmers and the tree farmers who "grow their crops together". One final advantage of the system is that it puts land to good use, providing not only a means of reforestation, but also helping to raise local incomes and to increase food production.

The only disadvantage that might be considered is the competition of the wheat and the tree seedlings for nutrients and soil moisture. However, in an area such as El Tollo with over 1800 mm of rainfall per year and on land that has lain fallow for 20 years, this should be of only minor consequence.

WIMCO, India

WIMCO Ltd also operates in a temperate zone—at locations both in Uttar Pradesh and elsewhere in northern India. The company raises fast-growing poplars as raw material for its mills, which produce matches, veneer, pulp and paper, fuel, lumber, pencils, packaging, joinery and light furniture. Having worked with farmers in the low- and medium-income groups for about ten years, WIMCO has been motivating landowners to plant poplars with wheat, lentils, sugar cane, potatoes and a wide variety of other crops. Once the trees are mature (in the eighth year), WIMCO buys the tree crop from the farmer at the market rate or at a predetermined price (whichever is greater). Through an extension programme the company supplies clonal planting stock and advises farmers as to how many rupees per hectare per year can be earned under different crop/tree combinations.

There are periodic visits by specialists who offer advice on planting and proper plantation maintenance. WIMCO also carries out an ongoing tree improvement effort, continuously introducing new varieties or provenances for testing. Trees are bred for narrow crowns and early leaf-fall to facilitate capture of sunlight by crops. This type of cooperation with local landowners has generated an immense amount of good will for the company and has significantly increased the income level of local farmers.

PICOP, Philippines

On the tropical island of Mindanao, the Paper Industries Corporation of the Philippines (PICOP), in addition to company plantations, has arrangements with about 5300 farmers who have planted some 20000 ha of *Albizia falcataria*. PICOP supplies seedlings at cost, payable at harvest time, and provides technical advice. The company guarantees to purchase all the wood produced at prices that are competitive at the time of harvest.

Lands within 100 km of the mill are eligible for this rather unusual AF scheme. Farmers are in need of land to plant their crops. The company offers them land on the PICOP concession, with ten parcels of land for each family. One parcel is planted to trees each year and one is harvested each year, with a 9-year rotation.

The tenth parcel is used to grow food crops for the farming family. Each of the ten parcels is planted in rotation to crops, then to trees, every ten years. There is also the option of growing crops among the trees during the first year of the rotation. The harvested wood is used as mechanical and thermomechanical pulp.

Jari, Brazil

In the Amazon Basin less than one degree south of the equator, the Jari Florestal e Celulosa Company conducts a massive tree plantation-agriculture-pulp mill operation. The site receives 2200 mm of rainfall per year, is 250 m above sea-level at the highest point and originally had tall tropical hardwood forest and very few local inhabitants.

The first step of the operation involves logging the natural forest for merchantable timber. Most of the remaining trees are used as fuel for the mill. The site is then burned and planted to either *Gmelina arborea, Eucalyptus deglupta, E. urophylla* or *Pinus caribaea* var. *hondurensis,* depending on soil properties. This is a rare example of cutting the natural forest in order to plant trees, a practice not normally recommended, given the vast area of deforested land already available.

At Jari, various food crops have been and/or are still being planted in conjunction with plantations of these four tree species. On sandy pine sites, grasses or manioc have been sown. Maize and beans have been grown concurrently with *Gmelina* and eucalypt seedlings on sedimentary clay soils. On Devonian clay soils, maize, beans and manioc have been grown between *Gmelina* seedlings. The grass is used for grazing some 6000 head of cattle and the manioc and maize are fed to the company's swine herd, which varies from 1000 to 4000 animals. Maize yields are about 500 kg/ha without fertiliser.

Since the company operates in a zone where few people live and yet requires a large supply of labour, workers had to be brought in from other regions. This requires a food supply which is largely being met by company efforts. Large-scale AF is the technique by which Jari maintains its labour supply. This may or may not reduce site maintenance costs on the company's vast tree plantations, depending on soils and the number of livestock per ha, but it certainly makes for a more efficient use of the site.

Celulosa, Argentina

Celulosa Argentina has been engaged in tree planting, mostly *Pinus taeda* and *P. elliottii,* in Misiones, Argentina, for many years. On some of these plantations the company has practiced mechanised farming, with maize and other crops grown among the pine seedlings during the early years of the rotation. The landscape is rather flat and the alluvial soils are excellent for agricultural as well as for silvicultural activity in this temperate zone area. The advantages and disadvantages and the silvicultural techniques required have been described in some detail by Cozzo.

La Fosforera, Chile

The company known in Chile as La Fosforera produces matches, wood-based housing and other wood products. It operates its plantations of poplar for a raw material source. These plantations are located on excellent agricultural sites, with fertilisation and irrigation used to improve growth rates.

The trees are grown in association with cattle, which graze beneath the trees during the rotation and feed on the tree foliage after the trees are felled. Like WIMCO in India, La Fosforera is associated with the transnational Swedish Match Corporation and uses its global connections for tree improvement purposes.

New poplar varieties are brought in from around the world for improved production or to develop other traits deemed desirable. An article by Ragonese and Giocchini describes similar efforts by several Argentine companies to combine poplar cultivation with livestock production.

Ion Exchange, India

Agroforestry is the main activity of Ion Exchange (India) Ltd, which can be described as an integrated agroforestry operation. The company proposed to reduce unemployment by planting trees on degraded sites—trees which reduce erosion and improve watershed properties and agricultural productivity as well as provide raw materials.

The raw materials are used for forage, fuel and poles, and will lead to such activities as apiculture, sericulture, rabbit farming and the culture of mushrooms, earthworms and fish, as well as a wide variety of other commercial enterprises. The company is quite new and has concentrated so far mainly on research to make all of this feasible.

The approach is novel and may lead the way for similar efforts in other countries. If it works in the warm temperate zones of India, it may well offer promise in other such areas. Whether this young company succeeds or fails, the techniques proposed should be carefully examined.

Fiat Lux, Brazil

On the temperate zone lands of Empresa Fiat Lux in Imbituva, Paraná, an experiment was established to test the use of a *Pinus elliottii* plantation for pasture. On a 4-year-old, 84-ha plantation, with 3 x 3 m spacing, cattle were introduced. The animals spent their winters in a nearby natural forest and the rest of the year in the pine plantation.

After 23 months, there were no negative effects measured on the pines. Although the soil was somewhat compacted by the animals, the comparison of pastures vs. non-pastured plots showed tree growth and survival to be the same. During a 3-year period, over 100 trend of cattle were sold. Which greatly increased the cost-effectiveness of the pine plantation. At the same time, four species of pasture grass were tested under four levels of pine-shade intensity.

This kind of research is needed to advance the practice of industrial AF. The private sector will probably do its own research to resolve localised problems, but public sector research, extension work and incentives programmes would do a great deal to promote and encourage wider use of these land-use maximisation techniques.

All the companies mentioned above have been faced with problems that hindered production. Most of these have been problems that could be managed, reduced or resolved. One of the advantages of this level of AF is that industrialists are accustomed to resolving problems and engaging in efforts that pay their own way. Most private enterprises are quite good at this; those which are not soon disappear.

FOREST INDUSTRIES IN SOCIO-ECONOMIC DEVELOPMENT

The establishment of forest industries provides potential either for foreign exchange earnings or—if production is consumed within the country—foreign exchange savings. Developing countries in 1982 exported major forest products for a total value of US$7100 million, whereas their imports of the same products had a value of US$10100 million. The biggest developing country share of world exports was in industrial roundwood 45 percent—followed by wood-based panels with 35 percent. On the other hand, their share in exports of sawnwood was surprisingly low only 17 percent—and, as might be expected, only 7 percent in wood pulp and 3 percent in paper and paperboard.

The developing countries also had a substantial share in the imports of industrial roundwood and wood-based panels: 20 and 25 percent respectively. Furthermore, they accounted for 18 percent of the total world sawnwood imports and for 31 percent of the pulp, paper and paperboard imports.

If developing country exports of industrial roundwood were reduced and at least some of the resulting savings invested in the production of sawnwood for export, considerable earnings could be made through the increased value added. This would also release processing residues for use as fuel in industrial plants, thus contributing to savings in foreign exchange now spent on imported fossil fuels. Other possible uses for these residues could be as fuelwood for households, wood-based panels or for pulp and paper. In all these cases, the pressure on the forest to provide goods would be reduced. Because of the capital intensiveness and the small markets in most developing countries, there is a question as to whether a substantial change can be expected in the global import/export pattern of pulp and paper in developing countries. While some improvements can be expected in individual countries or groups of countries, the picture for most developing countries is likely to remain the same, at least in the short term. Regardless of where changes in the export/import pattern might be achieved, it could be mentioned, as one illustration of the possible implications, that if the value of exports from developing countries could be increased by one-third and the value of imports reduced in the same proportion, the net value of these changes would amount to about US$5000 million over one year, based on 1982 prices.

In the context of development of forest industries, both primary and secondary processing need to be taken into account to ensure the best possible developmental impact. According to the existing agreement between FAO and the United Nations Industrial Development Organization (UNIDO), FAO is responsible for the development of primary forest industries whereas the secondary forest industries such as furniture manufacture, joinery, prefabricated housing, etc, are the responsibility of UNIDO. It is obvious that cooperation is needed between the two organizations and this is continuously being given through the FAO/UNIDO Joint Working Group on Forest Industries.

Contributions to Employment

Forest-based enterprises provide direct and indirect employment in a variety of ways. In some countries forest industries make up nearly one-fifth of total manufacturing employment. Moreover, recent figures indicate that rural small-scale forest industries provide the principal employment for between 20 and 30 percent of the rural labour force in many developing countries. The proportion of rural cash income de rived from these enterprises can be very high, rising on occasion to over 50 percent. This is especially important for the poor and landless members of society who depend heavily on these enterprises and who may derive most of their income from them.

Forest industries are also vital in stabilizing income throughout the year, as they can frequently provide employment during periods when agricultural activities are reduced.

The employment provided can be an alternative to a shifting agrarian lifestyle which, when badly practiced—as often happens where there is endemic poverty or rapid population growth—is one of the major causes of the diminution of the world's forests. Bearing in mind that about 200 million people in the world earn their livelihood from shifting cultivation, the significance of this activity should not be overlooked. Moreover, because of the cultural and social implications associated with shifting cultivation, the problems it causes must be approached with care.

Choice of Technology

Some of the issues that could be considered are the following:

- for existing enterprises (particularly rural small-scale enterprises), an introduced upgrading of technology can result in increases both in productivity levels and also in the size and competitiveness of the enterprise, but only if the application of the enhanced technology is concomitant with the skills that can be transferred to the workers;
- when a new enterprise is being located in a particular rural area, it may be possible to choose technology that the local population can apply immediately or can be readily trained to do so. If not, the industry may have to bring in skilled workers from elsewhere as part of the work force; if the newcomers are sufficiently different in terms of race, culture and lifestyle from the local population, conflicts may arise and the enterprise founder because of labour problems;
- among the problems to be addressed, the choice of technology is, relatively speaking, of minor importance for the processing of non-wood products in general. Such operations are usually less sophisticated and less capital-intensive than those based on wood products. For this reason, the introduction of such industries might show immediate results toward alleviating poverty in rural areas;
- more sophisticated technology does not always require higher labour skills and may, therefore, for certain products and conditions, be more appropriate than simpler technology. For example, a highly automated paper mill will require lesser labour skills for the operation of the mill through the application of *technology* for paper-making than a simpler process involving the art of paper-making. However, such a choice sets high demands on the facilities available within the country for servicing. The drawback is that the modern technologically more advanced sector tends to create fewer jobs. This is because modern technology has been developed for conditions in industrialized countries; thus some adaptation to the conditions of developing countries needs to be made.

Rural small-scale forest industries provide the principal employment for between 20 and 30 percent of the rural labour force in many developing countries.

Sustainability of small-scale enterprises In spite of me importance of rural small-scale enterprises (SSEs), as indicated above, there often appears to be a bias against them on the part of governments, or at least a lack of institutional ability to promote them. Credit, for example, is more difficult to obtain and government services are, in effect, less available since the SSEs usually have less political appeal and prestige than large projects. They are also difficult to reach by government agencies.

Non-wood forest outputs A considerable contribution to employment and income can be generated from the non-wood-based products and services provided by the forest. A partial list of such products includes rattan, leaf products, bamboo, cork, honey, resins, tannins, fruits, mushrooms, nuts and wildlife for food and for trophy hunting. The services provided by the forest include protection of watersheds, recreation and tourism. Although these outputs can and should be considered along with wood products in an integrated approach to forest management and utilization, this is often not the case. One reason is that different (and sometimes competing) government ministries are responsible for different aspects of the forest. Several ministries may be involved, including natural resources, industrial development, tourism, environment, agriculture and even defence. Often it is a lack of coordination between the forestry service and the industries ministry which results in an uncoordinated approach to utilization of wood and non-wood products.

Generation of Economic Surplus

The primary contribution to economic development of any industry is to generate an economic surplus which can be invested in further growth. Therefore, forest industries, like any others, must be judged on their ability to generate an economic surplus. This is a primary objective of both rural small-scale enterprises and large-scale industries.

In the past some projects have been justified largely on the basis of national prestige or social impact but, lacking profitability, these have ultimately failed or required heavy subsidies which take resources away from new, more productive investments.

In the case of an integrated complex of wood industries, the lack of profitability of one component should be accepted only if the total complex is profitable. Even then this component can be accepted only if it produces goods which are a necessity for the country's economy.

Some of the issues relating to the generation of economic surpluses can be considered under two headings: external and internal.

International trade, by expanding markets and allowing countries to specialize in the production of goods in which they have a comparative advantage, provides an important *external* stimulus to economic development. The developing countries' share of international trade in forest products is relatively small but highly differentiated. Whereas exports of pulp, paper, and wood-based panels are small or even negligible, plywood, veneer and broad-leaved sawnwood are products in which developing countries have substantial and increasing portions of international trade.

Seven countries in the developing world now earn over US$250 million a year each from the export of forest products. Increased trade in forest products could play a significant role in relieving the serious debt burden of many developing countries, either through direct earning of foreign exchange or through savings in foreign exchange expenditures if domestic resources can be substituted for imported products. Most of the past and present wood supplies in developing countries originate in old-growth or primary tropical forests. However, some developing countries have recently developed well-designed and well-run plantations which supply exports as well as domestic demand. This is an area where continued growth is certainly possible.

Unfortunately, international trade in forest products is restricted by a variety of intentional measures designed to limit trade flows, including direct tariffs, non-tariff barriers and non-tariff measures. Although tariff barriers have been reduced in recent years in both developed and developing countries, the effect of these measures has been more or less cancelled by the introduction of non-tariff barriers, especially for, processed products like wood panels. These include: discretionary inspections, import licensing requirements, time-consuming customs formalities and quotas that restrict the volume of a particular commodity that can be imported. Non-tariff measures such as health and sanitary regulations, packaging regulations and grading can also have important restrictive effects on trade.

With respect to balance of payments, the import of paper products represents a major drain on foreign exchange for some countries. However, the expansion of domestic paper production capacity requires very large investments and is difficult to develop. Nevertheless, if they are managed efficiently and supported by effective marketing and trade policies, pulp and paper mills can make a major difference to a country's balance of payments position.

The *internal* or domestic market represents the first possibility for many developing countries to put their available forest resources to an industrial use. Sawmilling is an ideal industry for this market as it provides urgently needed timber for construction and allows the installation of production units at costs within the reach of countries possessing sufficient raw material.

One of the issues with regard to the generation of revenue from local markets is that much of the activity is carried out by non-local entrepreneurs (who may nevertheless be nationals). Many of the earnings, in these cases, are therefore not being returned to sustain the economic base for future expansion or to benefit the local people directly.

Marketing

The key to making a successful contribution by forest industries to socio-economic development often lies in effective marketing.Local enterprises in developing countries often have inadequate marketing capabilities while foreign transnationals have a well-established and efficient marketing network to sell their products on the same markets. Especially with regard to domestic marketing, developing countries should be as willing to invest in effective marketing facilities as they are to invest in production units. This may be essential to maximize benefits and to ensure competitiveness with imported goods.

Industries in developing countries need distribution channels and trade organizations and agencies which can assist them in marketing their products competitively on the world market. This also requires local personnel trained in marketing.

Because of the widely varying wood requirements in rural areas, marketing activities need to be combined with extension services in wood use and application. Local rural enterprises usually cannot provide essential know-how and technical services. Cooperative efforts between consumer and producer interests, supported by the respective development agencies, should in such cases complement marketing activities. With the establishment of rural cooperatives, marketing could be improved and more of the benefits and experience gained could be directly utilized by the producing community.

If the value of exports from developing countries could be increased by one-third and the value of imports reduced in the same proportion, the net value of these changes would amount to about US$ 5000 million over one year.

Provision of Inputs to other Sectors

Forest industries often make a vital local contribution to the economy by helping to develop the physical and social infrastructure of rural areas. In their infancy these industries use the existing infrastructure, such as rivers and coastal waters, to reach markets. However, as the quantity of goods flowing from remote areas increases, roads are built and other means of transport are designed.

This increased economic activity in a community can lead to the establishment of some basic local processing, such as a small sawmill. The output of this primary wood processing unit in turn provides the raw materials for further local processing into higher value added products with the provision of more employment. As industries become more established in an area, the communities become more highly developed. Electricity becomes available, housing is improved and schools, hospitals and other facilities are established. Finally, the industry contributes to reducing illiteracy by providing paper for education and books.

The introduction of forest industries into a community gives members of the local population the opportunity to improve their skills and thus increase their income. These people are generally employed in logging and transport activities since employment in a mill tends to be comparatively small. Additional skills can be nurtured in workshops and services developed around the forest industries.

Trade between the rural and urban sectors, such as the sale of roundwood logs from a logging company to a processing plant, also influences infrastructural development such as highways, railways, power lines and ports.

Another example of inputs provided to the overall economy by forest industries is the production of packaging materials for a country's agricultural industry. It is possible that some agricultural activities might become non-viable if it were necessary to import packaging materials rather than obtaining them within the country. Other forest-based inputs are materials for making farm implements. In ways such as these, forest products can con tribute to agricultural production and food security.

The provision of building materials for a country's construction and housing needs is another important economic stimulus. In practically all developing countries, construction is a major industry as a result of population growth and the large shift of people from the land into urban areas. In many developing countries the products of forest industries, such as lumber, plywood and particle board, often form the basic raw materials of construction, especially in rural areas. An added benefit in the case of sawmilling is that it contributes to the local construction industry using almost exclusively local resources.

Mechanical products industries, however, are often hampered by a lack of support from government institutions, by their frequent inability to get loans and financing from lending agencies and by a limited supply of skilled labour.

Recognition of the three distinct contributions that forest industries can make to socio-economic development—the creation of employment, the generation of economic surpluses and the provision of inputs to other sectors of the economy—will lead national decision-makers to have a greater appreciation of the role that forestry and forest industries can play in meeting other national goals. Halting the often socially disruptive process of urban migration, for example, depends upon finding reasons for rural people to remain in rural areas: new forest industries could provide such reasons.

Whether or not to make investments to stimulate development in a remote area of a particular country, either to alleviate poverty among the people living there or to harness its natural resources, is a decision that often hinges on whether economic activity can be sustained in that area: forest industries, because they utilize renewable natural resources, can provide such a guarantee. If the harnessing of new technologies is a national goal, and if biotechnology is one of the principal new technologies, the tropical forests, where most of the world's plant and animal species are found, are likely to be the sources of many new forest-based biotechnological industries.

In meeting these varied goals, it is important to re-emphasize the fact that small, medium and large industries all have important roles to play. Just as a national economy operates through a network of village, urban and regional economies, so can the forest resources of a nation be used to create a network of household-based, village-based, urban-based and region-based forest industries of a variety of sizes and types. The role of these industries in national social and economic development will be substantial.

COMPANY-COMMUNITY FORESTRY PARTNERSHIPS

Changes taking place in policy, markets and civil society are generating increasing interest in the prospects for private-sector partnerships with communities or individuals for production of forest goods and services. At the same time, in many countries around the world, forests in large blocks are decreasing and forest farms are consequently growing in importance, so an increasing number of forestry partnerships pertain to goods and services produced outside the forest. Two major trends are influencing and drawing attention to the collaboration of private companies and communities: political pressures for local control; and globalisation of markets, capital flows and technology.

On the one hand, increasing attention is being put on forestry as a tool for local empowerment, whereby previously disadvantaged communities and individuals benefit from taking effective control and responsibility for decision-making regarding their forest assets. On the other hand, this is often occurring in contexts where, through privatisation processes and growing use of market-based policy instruments, private-sector control of forest resources and land is increasing.

This section examines the relatively new, but growing, range of company-community relationships—out-grower schemes, joint ventures, other contracts and informal arrangements—and discusses their advantages and disadvantages in relation to trees outside forests. Some emerging lessons are identified on the potential of partnerships to deliver security of forest goods and services and the conditions under which good partnerships develop.

Motives for Partnerships

The terms "company", "community" and "partnership" have a range of meanings, and various influences drive companies and local people to seek productive relationships with one another. Companies, from powerful multinationals to small community enterprises, are all motivated by profit to some degree, and especially by the need for a secure operating environment. Some companies are also motivated by concern for the environment and desire to make positive contributions to local and national development. Some are strongly influenced by environmental non-governmental organisations (NGOs) and by discriminating purchasing policies of buyers seeking products from well-managed forests.

Market and regulatory pressures are key incentives, while all companies are likely to want to keep the costs of landholding and labour down simply for reasons of good business. People at the local level often have considerable tree-management capacity because of their physical proximity to the trees, traditional or legal rights, indigenous knowledge and systems affecting land and trees, economic and cultural dependence on forest goods and services and sometimes well-developed mechanisms for resolving conflict. However, communities as such are rarely suitable units of social organisation for tree management.

Communities usually include many individuals and subgroups with widely different interests, rights, claims and aspirations for forest goods and services. For example, wealthier farmers may want to plant large numbers of one or two commercial tree species as a crop, while poorer farmers may want to plant various trees for diversification of products available to households. Before entering into partnerships, local actors may need to consider important trade-offs. For example, farmers considering involvement in schemes based on farm forestry must weigh the potential gain in wood, shade and shelter against the possible loss in agricultural production.

Types of Company-Community Partnership

The main formalised partnership arrangements are contractual out-grower schemes and joint ventures—legal contracts between two or more parties combining land, capital, management and market opportunities, formed with the intent to produce a commercial forest crop. There is wide variation in the nature of the relationship. For example, in some out-grower schemes the company pays the market price on delivery and exercises little control over production, while at the other extreme are schemes in which prices are fixed and the company exercises constant and rigorous control over all aspects of production.

The international pulp and paper companies Sappi and Mondi have been establishing out-grower schemes with farmers in South Africa since the early 1980s. In these schemes the company provides marketing and production services to farmers to grow trees on their own land under purchasing agreements laid out in a contract. Although they started as corporate social responsibility exercises, out-grower schemes are good business for Sappi and Mondi, even though the fibre produced costs them more than the average mill transfer price.

Company managers may find the schemes attractive because there is no need to invest in developing the company's own forest assets, and the costs are dealt with in the profit and loss account rather than on the balance sheet, where forest assets can all too easily become expensive liabilities. Today Sappi's outgrown timber is sourced from an area of some 88 000 ha in Kwazulu-Natal which includes 11 000 ha owned by 8 000 black smallholders and the holdings of about 260 white farmers with 50 ha or more each.

Farmers sign a contract with Sappi which entitles them to free expertise, silvicultural training and seedlings, advance payment for work and a guaranteed market for their trees at current market prices. When the trees are ready for harvest, Sappi pays the out-growers the market value of the produce, less any advance payments. Out-growers earn about US$205 per hectare per year, which compares favourably with alternatives such as livestock grazing or sugar production.

In Australia, three different types of joint venture arrangement have contributed to the planting of some 82 900 ha, or 8 percent of the country's plantation estate, since the mid-1980s.

- In lease joint ventures, the farmer signs over the land in a lease to the industry. Such schemes are attractive to commercial farmers and smallholders, as regular payments are made and indexed over an agreed period. With annual lease payments ranging from US$90 to $170 per hectare per year, returns are considerably higher than in many neighbouring grazing enterprises.
- Cropshare joint ventures are those in which the landholder and industry or government partners contribute inputs and proportionally share returns at harvest, based on the market price. Cropshare schemes are often attractive for underutilised agricultural land—often with poor access and low productivity—which does not always suit industry needs.
- Market joint ventures guarantee a sale for the grower, usually based on the market price at the time of harvest. The grower is required to offer the industry partner the first option of purchase, but if a better price can be found, the grower may sell to another purchaser.

There are indications that the range, scale of operation and types of partnership between companies and communities are increasing. For example, in an International Institute for Environment and Development (IIED) global survey of forestry companies producing wood pulp, 60 percent of the responding companies reported that they source some of their produce from out-growers or are otherwise involved in extension of tree-growing packages to farmers.

A more recent global review of private-sector involvement in forestry revealed at least one example of a company-community initiative over the past decade in 57 percent of the 76 countries covered.

It is often of limited value to think of company-community partnerships with- out considering the multistakeholder context in which they operate. Other stakeholders may include the following.

- Central government often has the important role of providing the policy framework and influencing the markets that stimulate or constrain partnerships. Government agencies may see partnerships as a low-cost means to develop forestry, improve local incomes and avoid giving too much to the private sector.
- Local government may play an essential part in setting the local terms within which partnerships can operate and in organising and monitoring them.
- Forestry officers may participate in brokering, mediating and monitoring partnerships, although they may be effectively marginalised through direct dealings between companies and communities. Forestry officers may also monitor adherence to forest and environmental laws.
- Federations or associations of farmers or other community-level actors may be the direct partner with companies who have limited capacity to deal with many individuals, or they may negotiate deals on behalf of individuals.
- NGOs may have considerable skills in brokering partnerships, and may sometimes do so with the indirect blessing or direct financial support of companies, governments or donor agencies.
- Banks may help overcome the problem of the long time scales involved in tree growing by providing loans to cover establishment and main-tainance costs against the future ability of companies or community groups to repay. However, it is often difficult for individual farmers to secure such loans.

There is a relative paucity of development literature analysing company-community partnership arrangements. The analyses available generally focus on contractual out-grower schemes and joint ventures, e.g. in India, southern Brazil and the Philippines, South Africa and Australia. Many contextual factors are involved in every case. In India the development of contract schemes with farmers has been heavily influenced by the fact that the State owns 95 percent of forest land and there are ceiling restrictions on companies' ownership of land, while in South Africa the big companies were initially more interested in such schemes as a means of demonstrating their social responsibility.

Thus, direct comparisons between schemes and ventures are of limited value. However, much can be gained from reflection on the experiences of a variety of partnership arrangements.

A partnership allocates risk between a local group or individual producer, which takes on the risks of production, and a company or contractor, which takes on the risks of the market. The appropriate arrangement will vary according to the policy environment, market conditions, organisations, credit, tenure and land type in question. Even similar farms in the same area may demand different arrangements. Thus there is a need for agile management approaches to activities such as marketing and quality control. Yet current partnerships are often pursued through inflexible and formulaic schemes.

Partnership arrangements (like other forestry enterprises), with their long time frames, are vulnerable to market instability. Many schemes have failed because of inadequate analysis of the price margin flexibility of both communities and companies, or because of poor forecasting of the consequences of changes in market conditions, such as the international boom and bust cycle in the pulp and paper industry. However, some partnerships are able to contribute to creating more vibrant local and regional markets, and thrive as a consequence.

The relative strengths and bargaining positions of the partners are largely defined by the contract between them and the economic and policy context in which it is embedded. It is rare to find a true partnership of equals—and perceived inequalities limit the adoption of many schemes. Frequently, individual producers and local groups lack the organisation and skills to enter into equitable agreements, and most would benefit from support for stronger associations to improve their bargaining power. In addition, bargaining power often changes over time, and the strongest partnerships seem to be those that are regularly renegotiated.

The contribution of partnerships to equitable livelihoods varies. In South Africa, smallholder woodlots developed under out-grower schemes provide a useful addition to household incomes, are often owned and managed by women and do not usually divert land or labour from other activities. However, in India, where secure tenure to land and some proved capacity and bargaining power are often needed to join partnerships, contract farm forestry seems to be a strategy for medium- or large-scale farmers in commercial agricultural areas. Furthermore, there is evidence of absentee landlords favouring these farm forestry contracts and thereby pushing tenants off the land.

Partnerships that rely on production for a guaranteed buyer may be temporary. As farmers gradually acquire skills, planting material and understanding of the market and become adept at fitting tree crops in with their other activities, their need for a formal relationship with a company may decline. A number of companies have incurred significant losses as a result of farmers reneging on contracts and selling their produce on the open market. However, more astute companies have realised that securing the produce from the farmers involves both paying the market price and building the relationship. Interaction between the company and the producer throughout the growing cycle—to deal with farmers' expectations, technical needs and problems—is crucial.

The more successful out-grower schemes set up supply centres for farmers which provide nursery stock of high quality and are profit centres in themselves. Farmers often need sound advice on the best techniques and appropriate technology for growing trees, but sharing of technical knowledge is also a two-way process in a partnership.

Challenges

Companies often find it easier, legally and operationally, to deal with individual outgrowers rather than with broader community groups. Many companies have relatively little understanding of communities' social dynamics or lack the ability to help build community capacity so that the partner will be willing, motivated and knowledgeable and so that internal disputes can be resolved when they arise. In South Africa, the government encourages private-sector investment in forestry, but the private sector still perceives considerable risk in dealing with local communities. A Forestry Development Office has therefore been proposed to act as an intermediary between investors and communities.

Responsibilities for developing partnerships are often blurred, particularly between government and the private sector. Markets currently, and increasingly, reward the short-term behaviour that policies and laws permit. Governments will have a key role in helping partnerships realise their potential as a tool for improving forestry and local development, but government capacities for intervening to support partnerships are not generally well developed. Governments will therefore need to learn from other sectors and from civil society. NGOs can sometimes offer considerable brokering, mediating and management experience.

Most current schemes focus on production of fibre only. As demand is growing, attention is increasingly turning to arrangements for equity sharing in downstream processing and other broader joint ventures. Such arrangements may need, for example, to accept the value of community land as equity. Furthermore, in the longer term, to secure their production base, companies will need to have wider socio-economic aims, including the provision of goods and services other than fibre. Partnerships with communities to manage land for non-fibre forest goods and services will assume greater importance.

FUTURE OF AGROFORESTRY INDUSTRY

Conversion of forests to other land uses will no doubt continue, at disastrous rates in some areas, and at controllable rates in others. Whether such conversions are controlled or not depends largely on factors such as fast remedial action, national will, local commitment, technology transfer/international cooperation and incentives. Funding is an important limiting factor, but these other factors are much more important. A great deal can be done at low cost if the other factors are present.

Agroforestry activity will increase at all three levels, as will technology transfer and extension efforts to promote it. Worldwide there will be far more tree-planting projects than at present and each of them should be considered a possible opportunity for agroforestry. The private sector will often find large-scale planting projects more cost-effective if industrial agroforestry techniques are employed.

Applied agroforestry research will increase as problems occur. The most successful AF research projects will be tied to large-scale plantation projects for reasons of economy. Incentives programmes to promote AF will be common from both the public and private sectors. These will be of several types but will mainly involve the incentives of increased profits and lowered labour costs, as well as public sector offers of tax incentives and free or at-cost seedling stocks.

Finally, there will be breakthroughs in systems for managing the natural forests of the tropics, both humid or dry. Once natural forest is managed successfully for sustained yields of a variety of products there will be less tendency to destroy such forests and convert them to other land uses. However, there will still be a need to complement natural forest management with plantations on adjacent sites for intensive, high-yield management. Plantation management will be all the more successful when combined with agroforestry techniques wherever appropriate.

REFERENCES

Curtis, A and Race D, *1998 Links between Farm Forestry Growers and the Wood processing Industry. Lessons from the Green Triangle, Tasmania and Western Australia,* RIRDC publications no 98 /41

David, Soniia 1995. *What do farmers think? Farmer evaluations of hedgerow intercropping under semi-arid conditions.* Agroforestry Systems 32: 15-28.

Dwivedi, A.P. and Sharma, K.K. (1989). Agro forestry: its potential. Paper in Seminar on Eucalyptus, FRI, Dehradun.

Eastin, I. 1996. *The role of marketing in promoting a sustainable trade in forest products in tropical regions.* Paper presented at the FAO/UPLB Workshop on Forest Products Marketing, Los Banos, Laguna

Falconer, J. & J.E.M. Arnold. 1991. Household food security and forestry. *Community Forestry Note No. 1,* FAO, Rome.

15

Women's Role in Agroforestry

Forestry has traditionally been one of the professions in which men have been most firmly and exclusively entrenched. It is not long since forestry schools—and hence the profession itself—were, in many places, open only to men. Arguments that the nature of the work makes it unsuitable for a woman persisted in forestry long after they had disappeared in other "manly" professions. The situation is changing fast. Women now make up a substantial part of the forestry student body in many if not most countries. And women are now practicing as foresters in both public and private service in sufficient numbers to have largely dispelled the earlier myths—though as contributors to this issue point out, these myths still persist in some places.

But the principal issues that need to be addressed in any consideration of forestry and women go far deeper than the question of career and job opportunities in the forestry profession, important though that is. More fundamental are issues relating to forestry and forest products in the lives of rural women in developing countries. Three-quarters of all families in developing countries depend on wood in cooking their food—fuelwood which is used and collected mainly by women. Forests are the source of foods enriching a predominantly cereal diet with protein, vitamins and minerals; they also yield a wealth of products that women can gather and sell. Plants growing in the forest provide most of the available medicines. Trees give shade to the home, and their foliage feeds the small animals that women widely keep around the home as a further source of food and income. In many countries, women are also involved in the manufacture of minor forest products at home.

In agroforestry projects, the involvement of women is essential in project definition, design and implementation. Yet women still, for the most part, are being excluded from involvement. Authors Fortmann and Rocheleau, after looking at four myths that are responsible for the exclusion of women from project participation, describe the purposes and results of the Plan Sierra Development Project in the Dominican Republic. This project, among its other accomplishments, clearly demonstrated possibilities for the employment and training of women in agroforestry.

Despite the crucial role of women in the development process, development projects continue to be designed without consideration of their effect on women or of the role of women in their implementation. A study of 43 World Bank forestry projects, for instance, found that only eight made specific reference to women. Agroforestry projects are no exception.

In part this state of affairs is due to the masculine images the word "forestry" conjures up. The reality is, however, diametrically opposed to the image. Women have traditionally played important roles in agricultural production and in the use and management of trees. The importance of these roles is, however, often obscured by the prevailing myths about the roles and status of women. Four of these myths are worth discussing.

1. *Women as housewives.* The first myth is that women are just housewives. In many, if not most, rural societies they are, in fact, farmers; often they bear the major or sole responsibility for food production. Region by region, country by country, ethnic group by ethnic group, detailed studies have documented the point that women's labour and women's decision-making are absolutely crucial to agricultural production and development. Women are frequently responsible for small-stock husbandry and the feeding of larger livestock, particularly milk cows and calves. Thus, agroforestry projects that involve fodder trees, the servicing of crops by trees or intercropping of crops and trees must include women, since it is often women who grow the crops or care for the livestock that will be involved.

2. *Forest products as the domain of men.* A second myth is that only men are users of and responsible for trees. In fact, women are the primary users of forestry products such as fuelwood, wild foods and fodder. Wood et al. noted that women "are primarily responsible for wood collection and utilization and often the initial establishment and tending of the wood stock around the village". Women may also make different uses of forest products than men. The priorities of men and women in species preference for Indian afforestation programmes, for example, often conflict. Hoskins also contrasts the interests of men and women in forest resources. Men are more likely to be interested in forest products for commercial sale and in products with end-uses farther from home.

3. *Men as heads of households.* A third myth is that every woman has a husband or is a member of a male-headed household. In fact, increasing numbers of women head their own households. Sometimes this is by choice. Sometimes it is the result of personal events such as death of a spouse, divorce, desertion or abandonment, or

of social trends such as male outmigration. Whatever the cause, woman-headed households are found in substantial numbers in every region of the world. It has been estimated that between 25 and 33 percent of all households in the world are de facto headed by wome. In short, the woman-headed household is an increasingly common social unit. And, as a result of heading their own households, women have assumed new roles. A historical process of women undertaking "male" tasks and working in "male" sectors in the absence of men has been reported for a number of societies.

4. *Women as passive community members.* A fourth myth is that women are passive members of communities in which public influence and public action is a sphere restricted to men. March and Taqqu have documented the important influence of women's informal associations in both the private and the public sphere. Within the domestic sphere, women exercise influence on public events through their information links, which are based on lineage ties, and on their ability to withhold goods (food, beer) necessary for men's public participation.

   Women's solidarity groups may take a very active stance in defending their own interests. For example, in the famous Nigerian "Women's War" of 1929, women, incensed over the rumour that an ongoing census meant that they were about to be taxed, refused to allow census takers to count them or their property. Women organized confrontations with the colonial administration that at times led to the use of armed force.

Women can gain leverage in the economic sphere through work and credit associations. They may influence events through their roles as healers or religious figures. They also participate in leadership roles in formal institutions. In Botswana they are generally the mainstay of social welfare organizations and even hold positions of authority in "male" organizations such as farmers' committees.

The influence of women and women's groups is demonstrated by a project in the Republic of Cameroon. Here, the men, fearing that the Government was trying to seize their land by planting trees, tore down a protective fence around the nursery. It was the women, recognizing the need for fuelwood, who persuaded them to rebuild it.

Women, then, both individually and in groups, have private influence on public action by men and undertake public action themselves. The potential for women's public action in areas such as reforestation and soil and water conservation is especially high because they are the principal sufferers from environmental degradation. It is they who must walk farther for water, fuelwood and fodder. It is they who must produce subsistence crops on increasingly degraded soils. It is they who often are both able and likely to organize the community for action.

WOMEN'S INVOLVEMENT IN AGROFORESTRY PROJECT

A case study of the Plan Sierra Development Project in the Dominican Republic can be used to examine women's involvement in different aspects of agroforestry. Plan Sierra is an integrated rural development project with strong agroforestry and reforestation components. During its first three years (1979-81) Plan Sierra developed innovative approaches to agroforestry, soil conservation, and forestry training and extension. While women were consciously included in some aspects of these programmes, their interests in "invisible" subsistence activities were nonetheless overlooked. Securing women's participation and serving women's interests were sometimes confounded. Both the successes and the failures of Plan Sierra's early years exemplify many of the key issues for women in agroforestry development projects.

The Sierra is a rugged, relatively isolated region in the Central Mountains of the Dominican Republic. Household income, health status and educational level of residents are well below national averages. The economy has suffered from boom-bust cycles in mining and lumbering. Out-migration has been high, particularly among men 20-40 years of age. The area has been largely deforested through commercial timber exploitation and the practice of shifting cultivation, which continues on the forest fringes. Soil degradation and erosion are widespread, and the region's hydrologic balance has been severely disrupted.

Most agricultural production is nested within agroforestry systems based on mixtures of field crops, coffee, pasture, and forest. These are combined in simultaneous intercropping mixtures as well as in rotations over time. Most farm families manage such diverse holdings for both subsistence and commercial ends.

Coffee and cattle are the major commercial enterprises among large landowners, while smallholders sell coffee or annual crops for cash. Most smallholder households are also heavily dependent on off-farm employment. Fuelwood and water are usually also gathered off-farm. Subsistence and cash crops alike are inter-cropped. Even pastures are usually studded with multi-purpose palm trees. Local cottage industries include furniture-making, food-processing (cassava bread, cheese, candy) and production of palm-fibre containers for sale to the tobacco industry and for local use with pack animals.

Women share with men the harvesting of annual crops and the coffee harvest (as owners and/or as hired farmworkers). Women raise the small animals (hogs and chickens) for meat and egg production, they usually milk the cows (for home consumption) and they tend patio (dooryard) gardens with vegetables, bananas and herbs. Responsibility for fuelwood- and water-gathering falls mostly on the women, with some help from the children.

As in many similar projects in Latin America, local women's participation in Plan Sierra was initially limited to health services, home hygiene and home economics. Cheese-, candy- and cassava-processing are almost entirely women's enterprises. Palm-fibre containers are produced and sold by women with help from children and elderly family members. Women artisans also weave the palm-fibre seats and backs onto locally manufactured wooden chair frames. The weaving is subcontracted as piece-work from men's woodworking shops.

Programmes

Plan Sierra was designed as an integrated rural development project to serve a 2500-km^2 area. It included strong agricultural, reforestation and soil conservation components from the outset. The programme of work was divided among the technical unit (agriculture, forestry), health, education and rural organization; surveying, engineering and construction; and marketing and rural industry.

Agroforestry and forest production were included in the programmes of the technical unit. Agroforestry initiatives emphasized coffee systems. Farmers (mostly smallholders) received subsidized credit and intensive training courses to facilitate the establishment of multi-purpose shaded coffee stands. Credit and technical assistance were also provided for the establishment or improvement of fruit orchards, often inter-cropped with annual food crops.

Aside from the planting of coffee and fruit trees, tree-planting on small farms was not treated separately from the reforestation of large-scale state and private holdings. Reforestation efforts focused on indigenous and exotic pine trees for watershed management or timber. These pine species were unsuited for small-farm use, given the historical circumstances of the Sierra, where the prohibition against cutting trees (regardless of land title) has been most strongly enforced in the case of pines and other timber species. Thus, most small farmers were unwilling to plant them on their own property.

Support services for agroforestry and reforestation included soil conservation, nursery development, employee training and community education programmes. Hundreds of local men were hired to construct nurseries and to engage in soil conservation and forestry training, extension, and construction activities throughout the 2500-km^2 project area. Technical unit personnel also promoted agroforestry and reforestation through teacher training courses in local ecology, reforestation and sustainable production systems. The teachers were later enlisted as participants and promoters in community tree-planting campaigns.

Women's Role in Agroforestry

As in many similar projects in Latin America, local women's participation in Plan Sierra was initially limited to health services, home hygiene and home economics. Women professionals were concentrated primarily in health, education and rural organization. The maturation of Plan Sierra programmes and their respective inclusion or exclusion of women provide some insight into issues that need to be addressed in future agroforestry and development projects in Latin America. The key concerns identified include access to employment, training, credit, land and appropriate technical assistance.

Local women were initially hired as home economists, secretaries, cooks and cleaners. Eventually some of the nurseries hired women to water seedlings and to fill polyethylene bags with potting soil. Nursery supervisors considered women to be more efficient and patient at this tedious task.

Some women, who began working only reluctantly with plant propagation techniques, later acquired an active interest in such work. Plan Sierra administrators and some technical staff encouraged this trend. They trained a group of paratechnical women horticulturists who became known as the "budders and grafters". The job attained a high status and was accepted as a women's task, setting a precedent for the inclusion of more women in technical nursery work. As in many other, similar projects, sexual harassment by a few co-workers and managers made work difficult for some women employees, but administrators eventually ousted the offenders.

Women began to request specific assignment to the nurseries, and a few expressed interest in conducting forestry or soil conservation fieldwork with visiting women researchers. Some of the "budders and grafters" talked of investing part of their earnings in land or small citrus groves, an option they would not have considered previously.

Technical assistance and training. Despite a high incidence of women heads of household, women were rarely direct beneficiaries of community or farm-level tree-planting. They were actively sought as volunteer labourers and promoters for community reforestation and also became involved in technical training and project implementation through the teachers' training courses. About half the teachers in the Sierra are women, and all teachers attend the same training sessions, regardless of individual specialization or prior training.

Women were seldom direct beneficiaries of technical programmes—not by design but by default. Since poor farm women were not consulted on technical project design, their concerns and needs were not adequately addressed in the technical programmes. When consulted during a mid-project evaluation of women's needs, women requested assistance with the development of water supplies, fuelwood supplies, patio gardens and cottage industries such as weaving and food-processing. Here, both fuelwood and cottage industry issues fell within the domain of agroforestry.

Some of the women who produced woven containers complained of a lack of access, or dwindling, insecure supplies of palm fibre. The palms are also important sources of wood, food and animal feed, and they are often located in pastures or fallow land owned by neighbours or relatives. The women have free access (an apparent advantage) but no guarantee of future access and no control over cutting and replacement. The palm-frond supply is both free and unreliable. Supplies in some instances were threatened by the felling of palms for cheap construction material. Many local men decided to use the trees for construction or cash income after their hog-feed value was undermined by a swine-fever epidemic and a subsequent embargo on hog-raising.

The management and improvement of this disrupted multiple-use agroforestry system were not integrated into either the pasture management or the rural industry programmes. The marketing of the finished products, however, was addressed by Plan Sierra (marketing and rural industry unit), with the farm-gate price of containers nearly doubling. Neglect of the technical and management aspects of this enterprise may have been conditioned, in part, by the low cash income from (and low priority given to) such enterprises as coffee. However, the importance of this cottage industry to rural women (especially the poor) lies in the fact that it requires neither land nor capital and that work can be performed by women, children and the elderly.

The fuelwood shortage also had the greatest impact on poor families with smallholdings. Some women had closed down their cassava-bread-processing operation because of the lack of fuelwood nearby and the high cost of purchased fuelwood. Others commented on the increasing time and effort (and trespass) required to secure the same quantity of fuel for home use.

Gathering, selection, consumption and the potential for subsistence production of fuelwood in rural farms or towns were not directly addressed in any of the programmes. Research on commercial charcoal production in dry forests had been conducted in the region but these studies were not utilized during this period. Some trials of exotic species, including fuelwood species, were initiated in 1981, but no experiments or farm trials were conducted with proven indigenous fuelwood trees.

The fuelwood problem was not initially recognized as a high-priority issue. When the issue did arise, there was a lack of training and experience among technical staff in the choice, propagation, management and promotion of fuelwood species, particularly those suitable for planting on the farm. Nor were there any women foresters (or agroforesters) available within the country. Thus, a major opportunity to involve and serve local women in agroforestry and reforestation was lost.

Women were likely to benefit directly from the technical assistance when their cash income or food sources coincided with those most often managed by men. For example, those who had (or wished to establish) coffee holdings, as heads of household or as spouses of owners or managers, were invited to attend training courses for improved agroforestry systems based on coffee. Women were consciously included, although they were not considered to be the main clients of this training/extension/credit programme. They were, in effect, given equal access to training and equal formal commitments for technical assistance in a field considered to be a man's domain.

Obtaining credit proved to be more difficult, and depended chiefly upon marital status. Some subsidized credit and land reform programmes were included within the overall project, but lack of access to credit and land constituted a serious obstacle to the implementation of on-farm or community tree-planting projects by and for women. Credit was particularly difficult to obtain for women settled in small towns, on non-titled land or on the property of absentee husbands.

Plan Sierra illustrates both the possibility for the employment and training of women and the need for further utilizing their experience in project definition, technology design, and extension and implementation. It demonstrates that even in regions where women do not traditionally till the soil, they can and should still be offered employment and training in nursery and horticultural techniques, some of which can be identified as "women's occupations".

The experience of Plan Sierra also indicates the need for prior consultation with women clients of agroforestry development projects about issues of immediate concern to them and about potential action to solve problems or otherwise improve their lot. This would, in many cases, imply a reordering of priorities in project identification, technology design and species selection criteria to better meet the needs of rural women, especially the poor and heads of households. The employment and training of women in fields already recognized as important (but not exclusively defined as men's work) could then be extended to the training of women personnel for more flexible roles in agroforestry extension programmes for rural families, including subsistence farmers and smallholders.

WOMEN'S ROLE IN THE CHIPKO MOVEMENT

The Chipko Movement has attracted world-wide attention. The image of poor, rural women in the hills of northern India standing with their arms around trees to prevent them being cut down is a romantic and compelling one. The reality, in many ways, fits the image: the Chipko Movement can indeed be considered an important success story in the fight to secure women's rights, in the process of local community development through forestry and in environmental protection. But there are more complicated implications as well. It is important to understand the history of Chipko and the context in which it arose—and is still evolving.

In the case of women's role in the Chipko Movement, it is both. (Chipko, a Hindi word meaning "hugging", is used to describe the movement because local village women literally "hugged" trees, interposing their bodies between the trees and the loggers to prevent their being cut down.) The Chipko Movement is an ecological movement, concerned with the preservation of forests and thereby with the maintenance of the traditional ecological balance in the sub-Himalayan region, where hill people have traditionally enjoyed a positive relationship with their environment. Thus, it strives to maintain the traditional status quo between the people and the environment. Its proponents have tried to demonstrate that the past and present forest policies of the Indian Government have negatively affected the ecological balance of the area and caused the uprooting of indigenous people who previously depended on forest for their survival and who preserved the forest by maintaining a strong bond of veneration and love toward it.

The Chipko Movement, which has now spread from one end of the Himalayas in Kashmir to the other in Arunachal Pradesh, is endeavouring to alter the Government's forest policy by insisting on maintenance of the traditional status quo in the Himalayan and other forest regions of India. In this sense, there is resistance to change and to an opening up of the area for technological development.

The collective mobilization of women for the cause of preserving forests has brought about a situation of conflict regarding their own status in society. Women have demanded to share in the decision-making process along with men; hence, there has been opposition by men to women's involvement in the Chipko Movement. Women are, on the one hand, seeking alterations in their position in society and, on the other hand, supporting a social movement that is resisting change. To understand this, it is crucial to ask why women support the movement, what the extent of their awareness is, and how many women in the hill areas are actually participating in the movement.

Women and Chipko. Leaders of the Indian independence movement at one stage decided to seek women's participation, and Mahatma Gandhi gave a call to Indian women to come out of their homes to work for the cause. In the Chipko Movement, women became involved through a different process. There was a sustained dialogue between the Chipko workers (originally', men) and the victims of the environmental disasters in the hill areas of Garhwal (chiefly women). Women, being solely in charge of cultivation, livestock and children, lost all they had because of recurring floods and landslides. The message of the Chipko workers made a direct appeal to them. They were able to perceive the link between their victimization and the denuding of mountain slopes by commercial interests. Thus, sheer survival made women support the movement.

Why men did not see these connections and women did has to do with the way the subsistence economy is organized in this area. It is also related to the way men perceive the Chipko Movement as a "back-to-nature" strategy and to their preference for a traditional type of economic development that takes place around them.

However, whether the Chipko workers realized it or not—or intended it or not—the women who participated in the Chipko meetings, processions and other programmes have become aware of their potentialities and are now demanding a share in the decision-making process at the community level.

Social Setting

The Garhwal division of Uttar Pradesh (one of India's northern states) comprises the four districts of Uttarkashi, Chamoli, Tehri and Pauri and covers a total area of 27 002 km^2, with a population of more than 70 0000 persons, less than 1 percent of the total population of the state. Uttarkashi and Chamoli, both border districts having the Indo-Tibetan boundary to the north, are the least-populated districts of the state.

The Indian Social Institute of New Delhi financed a two-month study visit to the Chamoli district by a group including the author in September-October 1982. Chamoli was selected as our unit of investigation because the Chipko Movement, initiated by a group of Sarvodaya workers (followers of Mahatma Gandhi's disciple Vinoba Bhave), originated here. The total area of the Chamoli district is 9 125 km^2. Ninety-six percent of the district population lives in villages. There are 1649 villages in all, and of these 1488 are inhabited. Of the total population, 58 percent are gainfully employed. Sixty percent of the total female population of the district are "working" while only 55 percent of the men in the district work. Further, 97 percent of working women are engaged in cultivation, as compared with only 72 percent of the men.

Not only do females in the Chamoli district outnumber males by four percentage points, but the single-member female households outnumber single-member male households. The majority in these single-member households belong to the 50-plus age group. Male migration from the hill areas to find work in the armed services and other jobs in the plains is fairly common, with women left to look after land, livestock and families.

A visit to the area makes one realize that topographic and climatic conditions require special adaptation by people who have to work extra hard to survive. During the 1982 field trip, seven villages were visited and open-ended interviews held with rural women and men. Unlike that of the villages in the Indo-Gangetic plains, the rural population of this area depends on land as well as forest for its subsistence and other survival requirements. Such dependence makes the character of social life in this region significantly different from that of the rural population in the plains. Nearly every family in the village owns land, usually less than half a hectare. Annual crops grown here are wheat, paddy, pulses and oil seeds. Farming is mainly dependent on monsoon rains rather than irrigation channels.

In general, subsistence farming by an average family of five members is possible for three to six months per year. For the rest of the year, villagers have to look for other sources of subsistence. The nearest source is the forest around them. Thus, settled agriculture is coupled with the foraging of minor forest produce. The villagers also use wood from the forest for various purposes, such as agricultural tools, dwellings, cooking fuel and fodder for grazing cattle. The use of forest products is expected to increase.

The women who participated in the Chipko Movement have become aware of their potentialities and are now demanding a share in the decision-making process at the community level. They asked the women to confine themselves to their fields and homes and simultaneously issued a written warning to Bhatt that if he tried to agitate or organize the village women, he would be killed.

People generally had free access to the forest until 1821, when there began a process of gradual control over the forest area by the Government. Among some nomadic tribal groups, control over territories holding strategic food resources was specified in terms of customary laws, but government policy specified their dissociation "from the management and exploitation of the forest wealth".

In terms of day-to-day life, the basis for sex-role differentiation and the types of relationship between the sexes are linked with the pattern of cultivation and exploitation of forest wealth. Women's position in the society is governed by the norms of a patriarchal system of social organization. Typically, men must prepare the land for cultivation because there are taboos associated with women operating the plough. Thus, women are never themselves able to initiate the process of cultivating; they must depend on men. Men also own the land, as property among the Hindus of Garhwal is transmitted patrilineally. The labour required to raise crops is, however, almost entirely supplied by women. Women do the planting, weeding and harvesting. There are no "prestige crops", raised exclusively by either sex. Most staple crops are raised by women, provided that men prepare the land by ploughing it for two days in each cropping season.

Government Programmes

The various development plans and tribal welfare schemes introduced by the Government have failed to make an impact either on the low standard of living in general or on the worsening conditions of women's household drudgery in particular. On the other hand, there are very visible signs of government-initiated development programmes such as those for road construction and the increased number of educational, medical and housing facilities.

Conversations with local teachers and students gave the impression that development in the forms of roads, schools, hospitals, hotels, shops, cinemas, radio and libraries had ensured increased participation on the part of the Garhwal region in the mainstream of national development. One old man stated in a calm voice: "Whether we like it or not, the government is opening up this area. For sure, the government is only working in its own selfish interests, and it has no aim of benefiting the people. All the same, it is up to us to benefit from the new developments, and if we want to take advantage of the new schemes we must prepare our selves to come forward and push the outsiders out."

Origins of Chipko Movement

The Chipko is one of many "people's" ecological movements that have sprung into being over the past 10 to 20 years. These movements are fundamentally different from ecological movements in the industrialized world. There, industrial pollution and even "development" are seen as threats, but threats primarily to present lifestyles. In the Chipko Movement, however, the basic concern is the very survival of the people in the hill areas. Rather than using the media to try to influence government policies, the people here have had to resort to a popular struggle.

Although the Chipko Movement was officially begun on 24 April 1973 by some Sarvodaya workers (all male) at Mandal, Chamoli district, the organizers had already been active in the field of social reconstruction for the previous 13 years.

One of the movement's leaders, C.P. Bhatt, and his co-workers, who belong to Chamoli district and who had worked for increased employment for local people, believe in the ideology of non-violence as propagated by Mahatma Gandhi and Vinoba Bhave. In 1960, they founded a workers' cooperative which organized unskilled and semi-skilled construction workers. For some time, they worked successfully in this field. One of their schemes, begun in 1964, aimed at creating more employment through the exploitation of the forests. The group established the Dasholi Gram Swarajya Mandal (DGSM) workers' cooperative and entered the market by buying forest rights through auctions to supply its small workshop making farm tools for local use. After initial success, however, the group was out-maneuvered by other, richer contractors. One Chipko leader said that, at present, 90 percent of the women are with him while 90 percent of the men oppose him.

In the meantime, the DGSM thought of starting a new enterprise—the collection of roots and herbs from the forest. In this activity, the cooperative gave employment to about 1000 persons between 1969 and 1972. In 1971 it opened up a small processing plant in Gopeshwar, which manufactured turpentine and resin from pine sap. Again the DGSM had difficulties, this time because the Forest Department did not allot adequate supplies of pine sap even when the price paid for it was higher than that paid by a partly state-owned producer in the plains. For eight months in 1971-72, the plant had to be closed down for lack of raw material. The plant therefore worked for a total of only four months. The Sarvodaya workers thus faced difficulties with government policies in each of their enterprises.

On 22 October 1971, villagers from nearby areas demonstrated in Gopeshwar against government forest policy. Meanwhile, the Forest Department, which had earlier refused the DGSM's annual request for 10 ash trees for its farm-tools workshop, allotted 300 ash trees to the Simon Company, a sporting-goods manufacturer from the plains, thus putting tennis rackets before the plough. In March 1973, the agents of the Simon Company arrived in Gopeshwar to supervise the cutting of the trees. There also arrived the Chipko Movement.

On 27 March 1973 at a meeting in Gopeshwar, local people decided not to allow a single tree to be felled by the Simon Company. A month later, DGSM workers and villagers from nearby areas marched out of Gopeshwar to Mandal, beating the drum and singing traditional songs. It was a rally of about 100 persons. The Simon Company agents and their men retreated from Mandal without felling a single tree. This event had an impact on the Forest Department, which now offered to let the DGSM have one ash tree if it allowed the Simon Company its full quota. The DGSM refused and the Forest Department increased its offer to two, then three, five and ten trees—the DGSM's original request. Finally, the Forest Department had to cancel the Simon Company's permit and the trees were assigned to the DGSM instead.

The Forest Department also ended the ban on pine sap supplies, but at the same time it allotted the Simon Company a new set of ash trees in the Phata forest in another part of the district. On 20 June 1973, a local leader joined hands with the Savodaya workers and organized a Chipko demonstration in Phata, 80 km away from Gopeshwar. Villagers of Phata and Tarsali kept a vigil on their trees until December, thus starting the long story of the Chipko Movement.

Thus far the movement had confined itself to the problems of unemployment among the local people. Earlier, the Sarvodaya workers had organized them in several enterprises. Among these activities was a 1970 relief operation, started when monsoon rains flooded the Alaknanda river and swept away hundreds of homes. During the operation, the workers realized that the chief cause of the flood was soil erosion from the clear-cutting of mountain slopes by the lumber companies. Despite the Forest Department's policy of planting cleared slopes, the base slopes remained bare. Overgrazing and gathering by villagers also caused the baring of many slopes. Another cause of landslides, the DGSM workers pointed out, was road-building.

In 1973, monsoon rains again brought a spate of floods in the area. By this time, the DGSM had fairly well spelt out its interconnected goals of raising local people's consciousness about the Government's forest policy, about their rights to use the local forest and about their responsibility to preserve the environment through a programme of afforestation. During the 1973 flood-relief operations, the DGSM workers observed the sad plight of the women who had lost their houses, farm and cattle in floods. The series of recurring landslides that followed (1977, 1978, 1979) caused severe damage to life and property, making villagers almost paupers. Working in areas affected by floods and landslides, C.P. Bhatt and his companions heard long stories of suffering by women. This experience gave them both an insight into women's problems and an unprecedented direct contact with them.

Branches of Chipko

As the years have gone by, the Chipko Movement itself has acquired two distinct streams of thought, personified by its two leaders, C.P. Bhatt from Gopeshwar, who pioneered the movement, and Sundarlal Bahuguna from Silyara in the Tehri region. The operational style of these leaders is totally different. While Bhatt is a grass-roots worker and believes mainly in organizing the people, Bahuguna a journalist, is a publicist par excellence. Though Bahuguna has also organized some protest activities in his region—for instance, Chipko activists in Henwal Ghati once went to the forest to bandage wounded trees with mud and sacking to protest against the indiscriminate tapping of pine trees—his main focus has been on spreading the message of Chipko far and wide. In 1981, Bahuguna started on a foot march from Kashmir to Kohima to campaign against deforestation.

Bhatt, on the other hand, has dug deep roots in the Chamoli region. He is, as a result, far less well known than Bahuguna. Bhatt has realized that it the local village communities have the right to control their surrounding resources, they must also undertake to conserve and develop those resources. So he has organized the country's largest voluntary afforestation programme through eco-development camps. These camps bring together local villagers, students and social workers who have planted over a million trees. The survival rate of these Chipko plantations has been an astonishing 85 to 90 percent in most cases. Bahuguna, however, tends to dismiss this activity as irrelevant at this stage of the movement, concentrating all his writing and speaking power against the forest departments.

The two leaders differ not just In their operational styles but also in their philosophy with respect to the use of forests. Bahuguna is fiercely ecological in approach. The re-greening of forests is the top priority—a matter of national defence—for him. For instance, he argues that the main objective of forest management in Himachal Pradesh should be soil and water conservation: forests, he says, do not produce timber, resin and foreign exchange but soil, water and pure air. The self-sufficiency of the hill people in food, clothing and shelter is important to Bahuguna but secondary to the major ecological objectives.

For Bhatt, however, the search for a new eco-development process for the region and the involvement of the local people are primary issues. "Saving the trees is only the first step in the Chipko Movement," says Bhatt. "Saving ourselves is the real goal. Our future is tied up with them."

Bhatt, therefore, wants forest resources to be used in a manner that is both environmentally and developmentally sound—in other words, while the environment is preserved, the benefits of the controlled exploitation accrue to the local people, a process in which decentralized economic growth and ecological conservation go hand in hand.

Notwithstanding the divergent opinions of these leaders, the real strength of the movement is the women of the region. Except for a few "organized" events, the Chipko Movement essentially consists of a string of spontaneous confrontations in which none of the so-called leaders is present.

The Chipko Movement is thus very much a feminist movement. It not only has brought forth in a dramatic manner a greatly increased understanding of the divergent interests of local communities and state bureaucracies in the management of local resources; it is now finding that the interests of men and women within the same community can differ greatly. As long as the leadership of the Chipko Movement remains sensitive to this learning process, it is bound to grow in strength. The latest demand to emerge from the women of Chamoli is that it is they who should be elected to the Forest Panchayats and not their men.

Confrontations

When the Forest Department announced an auction of almost 2500 trees in the Reni forest overlooking the Alaknanda river, which had flooded in 1970, Bhatt reminded the villagers of the earlier flood and warned of more landslides and more floods if the remaining forests were cut down. He suggested that they hug the trees as a tactic to save them. Who listened to him? As subsequent events showed, it was women rather than men who got his message. One woman, Gaura Devi, organized the women of her village, Lata, and faced down the workmen of the company that had won the auction for felling the trees. It was a situation that almost forced women to take action—which they did with firmness and unyielding courage. Gaura Devi later described the encounter in graphic detail, commenting on the rude behaviour of some of the men and on how she pushed herself forward in front of the gun of one of these labourers. She challenged the man to shoot her instead of cutting down the trees, comparing the forest with her mother's home (maika). Eventually, she and her companions forced the men to retreat.

Following this demonstration of strength by women, the Uttar Pradesh Government decided to set up a committee of experts to investigate the situation, and the lumber company withdrew its men from Reni to wait for the committee's decision. The committee, after two years, reported that the Reni forest was an ecologically sensitive area and that no trees should be cut in this region. The Government placed a 10-year ban on all tree-felling in an area of over 1 150 km^2. This event blazed a trail: at Gopeshwar in June 1975, at Bhyndar valley ("valley of flowers") in January 1978, at Parsari (Joshimath) in August 1979, and at Dongri Paintoli in February 1980, women took the lead in Chipko demonstrations and saved forests from felling. After the Reni success, Bhatt and his workers began to address themselves to women and found them very sensitive and responsive to ecological problems. Women who were never before seen in any of the village meetings were asked to attend. They welcomed this opportunity and turned out in great numbers.

Implications for society

Political involvement. The events at Dongri Paintoli village, according to Bhatt, indicated a new development in the movement. During a meeting between the members (all male) of the village council and the officials of the Horticulture Department, it was decided that the oak forest near the village would be given to the Horticulture Department for felling. The department, in turn, would provide the villagers with a cement road, a secondary school, a new hospital and electricity for their village. Some DGSM workers, together with Bhatt, tried to explain the implications of development and the importance of conservation. However, the village men, especially the members of the village council, did not agree. They maintained that a school, a hospital, a road and electricity were far more important for the village than a few hundred trees. Yet the efforts of Bhatt and others did not go to waste on the local women, who decided to hold a Chipko demonstration if anyone tried to fell the trees. They even asked Bhatt and his men to help them. On hearing about this, the members and president of the village council became infuriated at the "outrageous" behaviour of their women. They asked the women to confine themselves to their fields and homes and simultaneously issued a written warning to Bhatt that if he tried to agitate or organize the village women, he would be killed upon arrival at the village.

All this did not deter the women of Dongri Paintoli, and on 9 February 1980 they did not even wait for Bhatt to arrive but turned out in large numbers, held a Chipko demonstration and prevented any tree-felling. Nine days later, the Government ordered the forest-felling in that area stopped, and within a month a ban on any further cutting was effected. Subsequently, women leaders in the village were defamed and asked not to attend further meetings. The women in Reni took action only because there were no men in the village around to do so. Their "action" was to ask the tree fellers to wait until their men returned so that some discussion could take place between the two sides (of men) as equals. Women took charge of the scene only in the absence of men, but once they did take charge, they succeeded.

In Dongri Paintoli, by contrast, rather than merely taking a decision in the absence of men, the women stood up against decisions made by their own men. Although they faced opposition from men, they held to their conviction. This certainly marked a major step forward in terms of women's role in the Chipko Movement.

In Gopeshwar, women have now formed a cooperative of their own, the Mahila Mandal, to ensure protection of the forest around the town. Its work is carried out regularly by watchwomen, who receive regular wages. Under their supervision, the extraction of forest produce for daily necessities is accomplished in a regulated manner, so as not to harm the trees. Women or men violating these rules are fined, and these fines are deposited in a common fund. Those who do not obey the rules face the punishment of having their tools confiscated. In addition, more and more of the DGSM educational camps are now attended by women, who come despite their busy routines. They take part in discussions and become articulate in expressing their views through this mode of informal education. Their programme, of course, is only in its initial stages. In most villages, women were found to be too busy in their day-to-day tasks to have time for the Chipko meetings and camps.

It can only be said that the cases of Reni and Dongri Paintoli and the organization of women into the Mahila Mandal at Gopeshwar are indicative of the latent potentialities in the organization and mobilization of resources by women whose consciousness has been raised. A situational analysis of the crisis periods shows how village women work in handling their problems: when new ideas and methods of handling problems are introduced by leaders, they are quick to act.

The situational conflicts in Chamoli district arose because of the different meanings attached to the word "development" by different groups of people. Men, who sit on village councils and other village bodies and head their families, view the government officials with a great deal of respect and fear. They dare not oppose them. Women, on the other hand, who have never had any contact with government officials or other outsiders, have no model of interaction to follow with them. The Chamoli women understood only that the felling of trees is harmful to their well-being, and they simply acted according to that belief. On the basis of their past interaction with government officials, men are convinced of the great powers of the Government. They consider it wrong to oppose its policies.

Women's participation in the Chipko Movement, however limited in numbers or in its impact on the general way of life, has implications for possible changes in gender relationships in the Garhwali society. One Chipko village leader summarized the present situation of the movement by saying that, at present, 90 percent of women and 10 percent of men are with him while 90 percent of men and 10 percent of women oppose him. He considers that only through non-violent methods will the movement win over the other men.

What we read about women's participation in the movement and what its leaders talk about are simplified and idealized images of reality. This idealization has, in turn, led to an unrealistic belief that the participation of women in the development process can be achieved by a mere ideological commitment and a few organizational devices. The account given here demonstrates that the release of spontaneity and creativity on the part of rural women in Garhwal is chiefly a byproduct of actions initiated at the grass-roots level by the Sarvodaya workers to increase people's awareness about the environment. At present, these workers and their leaders face the problem of handling an unforeseen release of womanpower in this area.

Ecological balance is an important aspect of new approaches to development, and women's concern with local ecological problems is vital. In a majority of existing programmes for women's development, the top-down approach is used. Decision-making, evaluation and control rest at the top with planners and policy makers, while participants lack the scope to develop their own skills or to have any political say in deciding their own affairs. If we aspire to change in the social and political situation of women, we have to look at alternative approaches to replace the traditional power structure; hence the need to study women's participation in social movements.

FUTURE CONCERNS

The women who worked as "budders and grafters"—as well as tree planters—in Plan Sierra benefited in terms of wages and knowledge, which they could transfer to their own private economic activity. However, the number of such women was, of necessity, limited by the size of the project. The project, however, did not address two of the issues most important to rural farm women: fibre for handicrafts and fuelwood. Unless care is taken, such women's participation in a project through their labour may give them relatively little return if their needs are not considered.

The division of labour in agriculture has often resulted in men's involvement with export cash crops and women's involvement in the subsistence sector. Thus, Plan Sierra focused on the men's export crop of coffee while the women's non-export cash enterprises and subsistence activities received lower priorities. There is a clear need to establish who uses what trees for what purposes before planning a project. Lack of access to credit and land constituted a serious obstacle to the implementation of on-farm or community tree-planting projects by and for women. Agroforestry projects that involve fodder trees, the servicing of crops by trees, or intercropping of crops and trees must include women.

Project design must take into consideration what resources women actually have to work with. Either resource constraints must be alleviated or a project must be geared to the resources women actually control. Failure to do this will result in an exclusion of women from project benefits. In terms of contact with extension workers, the case has been made again and again for female extension workers to work with women. This strategy not only avoids the difficulties of trying to restrict male-female interaction but is likely to facilitate communication between the female farmer and the extension worker. Alternatively, reluctance on the part of the male forestry officials or extension staff to work with women can be overcome by requiring them to report their visits with women.

The nature of women's participation in agriculture has been shown to differ with their social class and control of resources. Just as the priorities of men and women may differ, the priorities of rich and poor may differ. This point is especially important to remember when women's organizations (formal or informal) are involved in promoting or utilizing agroforestry. It is often the case that only wealthier, better-educated women have the leisure time to spend in formal organizations or in training to obtain skilled employment.

Such women also often capture the leadership of both formal and informal organizations. While there is a great deal to be said for utilizing the organizational capacity of women's organizations and associations, care must be taken that the poor are not excluded. Associations should be identified that are "structured so as to redistribute introduced resources equitably" and in which "all members participate equally in or have equal access to group decision-making procedures and avenues of redress"

Women are traditionally the prime participants in both the agricultural and the forestry components of agroforestry production systems. They are also private and public participants in community life and decision-making. The failure to include them in agroforestry projects has several detrimental effects. It excludes from project benefits the increasing proportion of rural households headed by women. It may prevent project designers from benefiting from women's special knowledge. It may exclude (or even harm) activities and commodities such as fuelwood, basket-making and minor forest products, which are part of women's economic sphere. The participation of women is essential for the success of agroforestry projects, but such participation may, in the long run, also require changes both in approach and in the nature of personnel in forestry and extension departments.

REFERENCES

Joshi, G. 1981. "Forest policy and tribal development". *Social Action*. 31: 446-468.

Lancaster, C.S. 1976. "Women, horticulture and society in sub-Saharan Africa". *Am. Anthrop.*, 1978: 539-564.

ESCAP. 1980. *Report of the expert group meeting on women and forest industries.* Bangkok.

FAO. 1983. *Rural women, forest out puts and forestry projects,* by M. Hoskins. Rome.

World Bank. 1980. *Forestry projects and women,* by G. Scott. Washington, D.C.

16

Agroforestry Research

The fundamental constraint to effective agroforestry research in the developing countries has been a lack of political support, at both national and international levels, resulting from an insufficient understanding of the links between a strong national forestry research capacity and the resolution of broader development issues. This, in turn, has led to a long list of direct impediments: poor leadership, poor salaries, poor career prospects, poor training, poor funding and poor facilities.

In many developing countries expressions of political commitment to the forest sector are not reflected in institutional support. In fact, there sometimes appears to be an attempt to offset the lack of support to forestry institutions with vigorous declarations of intent. Within the sector as a whole, forestry research organizations tend to be particularly ill-equipped for meeting the demands placed on them with regard to helping forestry play its full part in addressing long-term socio-economic development challenges.

It would not be appropriate, however, to call into question the commitment of most developing countries to support forestry institutions because, without a laudable concern to give practical expression to the idea of national sovereignty, national forest research organizations would never have outlasted the independence process and emerged with a new identity. Without a strong sense of solidarity, the national, subregional and regional institutions responsible for forest research would have been dissolved long ago. For example, the establishment of the ministerial Conference for a Concerted Policy to Combat Desertification and Protect Nature (COMIDES) within countries affected by this phenomenon is a clear indication that our political leaders are keenly aware of the problem and are determined to confront it. The political support for the Tropical Forests Action Programme (TFAP) also underscores this.

The most recent campaign was engaged at the United Nations Conference on Environment and Development (UNCED), where the member nations of the Permanent

Interstate Committee for Drought Control in the Sahel (CILSS), supported first and foremost by the Group of 77, secured the adoption of a plan to negotiate a convention on desertification, thereby confirming their wish to relaunch a topic which was slipping down the political agenda.

The fact is that, in a period of general recession and scarce overall resources, the developing nations are obliged to implement "survival" policies (for example, concerning public health and education) while at the same time seeking to safeguard and exploit natural resources; this necessitates difficult choices. This having been said, however, given its multiple role and crucial importance in both everyday survival strategies and long-term conservation and development objectives, research into tree and forestry-related questions deserves unswerving commitment in which political patronage is backed by institutional support. Moreover, concern for national sovereignty in the establishment of research priorities would in itself justify a long-term financial commitment from the state.

FUNDS FOR RESEARCH

Funds for forestry research in the developing countries basically come from two sources: national coffers (through the national agricultural research system or the forestry service) and research components of external assistance efforts. With a top-level policy commitment' additional allocations should be made to forestry and more of the total should go to research.

The allocation to research of a larger percentage of total funding for forestry development is clearly in the interest of development projects (which are among the main users of research findings) in that it should enable them to have forward-looking research available rather than, as often occurs, research results that are only of use in emergencies. It is a disturbing truth that, in some countries, the budget and even the staffing of a single externally funded and implemented forestry development project are often larger than those of the whole national forestry research sector. For those of us with first-hand experience in forestry research in developing countries and who know the huge challenges facing the sector and the modest results that have been achieved, it is difficult to subscribe to the thesis that no more than 0.26 percent of forest production value should be earmarked for forestry research, as is the case in the developed countries.

A great strain is often placed on forestry research organizations when they have to seek bridging loans in order to continue work on plant material that has been abandoned because funding, either internal or from an externally funded project, has been terminated as a result of short-view planning. This is a major challenge but it is vital that information collection continues uninterrupted, as any break would jeopardize its accuracy. One way to guarantee the independence of and regular financing for research activities would be to create a forestry research fund, financed from development projects as well as with a portion of revenues gained from forest and forest resource users and/ or a substantial contribution from the national forestry fund. Since there is little scope for research to generate and plough back its own income in the short term, such a mechanism will initially need strong national backing. In this sector, as in many others, a major expression of gratitude will be owed to those who play a significant part in the removal of the main stumbling blocks.

In terms of external funding, while the tradition of bilateral cooperation between developed and developing nations is highly commendable, it should be noted that this form of technical assistance for forestry research is not always underpinned by financial support. If foreign funding is sought, it is desirable that the outside partners take over the commitments left unmet by the previous donors. The support of such partners will continue to be enlisted, in accordance with UNCED's message which urges us to consider the world as a global village.

To be effective, their contribution must be consolidated, coordinated and sustained. If the funding of a project based forestry research programme seems too risky, those concerned should not flinch from postponing it until a more favourable time rather than launching initiatives that are doomed to fail.

Donors could be asked to contribute to these resources in the context of negotiations for a common forestry research fund using unassigned credits. However, it should be borne in mind that, before entering into any commitments, many donors insist on having information about the background to the research programmes they are being asked to support as well as on knowing which countries are taking part.

A more palatable option from the donors' point of view might be an agreement covering all research cooperation in the sector and specifying all the commitments made by both sides, with a view to carrying through the planned activities as effectively as possible. It should be noted that, although the availability of regular funding would prevent data losses and an over-reliance on perceived trends, it obviously would not eliminate the need for full forestry trials.

INTERNATIONAL COOPERATION

The above-mentioned challenges facing national forestry research organizations militate in favour of collaborative efforts for the purpose of tackling issues of common interest. Each country may set up the bodies best suited to its forest policy aims and resources, but the time has now come to translate into action the decision-makers and political leaders' calls for collaboration.

This would provide an opportunity to extend forest research beyond the networks now being set up, by identifying "leader" countries to look after some of the concerns of the various members. While this would mean sacrifices on the part of those selected, these countries would also be contributing to the creation of centres of excellence in strategic research.

The priority aim would be to develop competent, motivated national teams of researchers, equipped with the requisite resources. Forest research facilities in developing countries are extremely fragmented and human, logistical and financial resources are widely dispersed. They need to be more concentrated in order to create viable long-term structures with efficient teams able to gear research to the level of the countries concerned.

The overlapping of substandard structures in the same ecological area is no longer a realistic option; the time has come for cost-effective and well-organized undertakings. Whether it is a matter of a programme, an operation or a single piece of research, a subject such as the genetic improvement of Faidherbia (Acacia) albida should not be handled by each interested country separately. Similarly, it now seems illogical for each country to set up its own research unit on the in vitro culture of ligneous plants.

However, at the same time a word of caution is merited against the unwarranted formation of too large a number of, or even competing, specialized regional support mechanisms. For example, despite the commendable political will that spawned the Sahara and Sahel Observatory, this body has not enjoyed the best of relations with the CILSS or the Intergovernmental Authority on Drought and Development (IGADD).

It is to be hoped that the Forestry Research Support Programme for Asia and the Pacific (FORSPA) can be extended and adapted to provide a model for similar networks in other regions and subregions. This would be a first important step in the organization of the sector and would obviate the need to approach the main forestry research partners individually.

The Centre for International Forestry Research (CIFOR) has the potential to be of significant benefit to the developing nations, particularly if it profits from the experience of the Special Programme for Developing Countries (SPCD) run by the International Union of Forest Research Organizations (IUFRO). In all cases, CIFOR must build on the progress already made, and support and complement the initiatives of such bodies as FAO and the International Center for Research in Agroforestry (ICRAF) to strengthen forestry research capacities. It must also devise mechanisms for the regular supply and exchange of plant material and scientific and technical data among forestry research bodies. These two areas, both vital to any form of research, are at present dealt with on an ad hoc basis.

At the international level, one can only deplore the eligibility conditions of the Global Environment Facility. If the developing countries are unable to convince their partners that environment, desertification and reafforestation are interlinked, it is vain to expect forestry research to be able to draw on these funds. Suffice it to note that the monitoring of permanent plots for the study of plant dynamics as part of biodiversity studies necessitates the long-term availability and mobilization of adequate financial resources.

The conservation and wise use of forest and tree resources has become an essential part of agricultural and overall development strategies. In this context the need for capacity-building in forestry research is recognized in Agenda 21 and the forestry principles, promulgated by UNCED. Sadly, a lack of long-term political commitment and the resultant lack of continuity in the provision of resources has made it impossible for forestry research to make its full contribution to development efforts.

Political leaders in the developing nations must be convinced that quality forestry research infrastructures and results are necessary. They must also understand that they require a sustained political and financial commitment as well as a skilled staff and that priorities need to be determined. The fact that the sector straddles other areas also requires an openness to other disciplines and thus a redistribution of resources that reflects these new horizons. Endorsing these principles by providing a steady injection of needed resources is the best way of persuading international partners to continue their assistance as well. The cost of programmes to ensure continued development of forestry research infrastructures in the developing countries may be high today, but if we fail to act promptly it may well become astronomical.

FARMER PARTICIPATION IN AGROFORESTRY RESEARCH

Generally, on-farm research ranges from statistically rigorous experiments to informal observational trials carried out by farmers, although some practitioners would even go so far as to include certain types of project monitoring and evaluation. As Hocking and Islam have observed, the essential feature of "research" in the broad sense is observation and the feedback of results to modify future action.

Different on-farm research efforts embody diverse styles of interaction between researchers and farmers. Ultimately, the style of interaction determines who "owns" the research. The chart describes four possible styles of interaction between researchers, on the one hand, and "farmers", on the other. There is some form of "participation" in all four interaction types which, however, are clearly different in the degree of local participation which they embrace.

Consultative Approach to Farmer Participation

During the 1980s, ICRAF researchers developed a "farming systems" approach to on-farm agroforestry research which was based on the Diagnostic and Design (D&D) methodology. This kind of consultative approach has resulted in a significantly increased participation of farmers in research with professional researchers. Nevertheless, ICRAF researchers discovered that the consultative paradigm embodies considerable limitations as an approach to developing truly adoptable agroforestry technologies.

In ICRAF's first on-farm trials, researchers tended to dismiss as unimportant such farmer-identified problems as "lack of shade trees" or "lack of fruit-trees" because, at the time, they did not see these as posing many interesting research questions. Of tar greater importance to the researchers was the diagnosed long-term decline in soil fertility; this affected the productivity of the whole land-use system and, not insignificantly, offered the researchers greater scope for interesting research.

Farmers concurred with the researchers on the identification of soil fertility as a problem but were sceptical that much could be done about it. So the soil fertility problem received a low-priority rating in the farmers' minds. What the farmers' sense of priorities called for was an extension solution, i.e. an immediate logistical response that would increase the local supply of seedlings of shade and fruit-trees.

In a more farmer-participatory approach, where farmers themselves set the priorities, these issues would have been addressed from the beginning, thus leaving the farmers free for a greater subsequent involvement in the more challenging research activities. Therefore, the early on-farm trials of alley cropping and other soil management technologies suffered from rather disinterested participation by the farmers. By failing to address what the farmers felt was most important to them, researchers ran the risk of failing to engage the farmers in a meaningful dialogue. It could be argued that every on-farm research programme needs to have at least a small extension component.

With the consultative approach, a lack of awareness of existing strategies, or an insufficient appreciation of the level of local commitment to existing strategies, is often among the main limitations to the researchers' grasp of the local situation. Even with the best intentions, it may be difficult to avoid.

On-farm Experiments

A sincere attempt to get local land users involved in the design activity is an explicit feature of the approach practised by the early ICRAF researchers. Although there is nothing that prevents a more participatory use of the D&D methodology, most of the early applications remained within the limitations of the consultative mode, since it was usually the researchers who made the final decisions about prototype technologies for the initial round of on-farm trials. This is not necessarily an inappropriate approach so long as the initial design is intended not as a final solution but rather as the opening statement in an extended dialogue in which farmers can express their own ideas through modifications of the prototype.

In the Machakos alley cropping example, the farmers introduced a modification using the contour hedgerows of leguminous trees for fodder rather than for green manure. In this form, the agroforestry system was well appreciated by the farmers. Moreover, the hedgerows retained at least part of their effectiveness as an erosion control measure, and soil fertility was improved in accordance with the farmers own "brown manure strategy by increasing the amount of manure that could be collected from the pen-fed livestock. It may he argued that this level of mismatch between researchers' and farmers' perceptions is too much, even at the start, and that it could be easily reduced if farmers were allowed to design their own technical solutions in the first place.

It can be countered, however, that minimising the researchers' own design input is not necessarily conducive to the best outcome for the farmers. In the alley cropping example, the local farmers might never have thought of trying the innovation of fodder hedgerows on cropland it researchers had not demonstrated it to them.

"Researchers have a tendency to be arrogant about their abilities and imagine that, alter the D&D and list of ''agroforestry interventions" have been identified, they have all the answers and all that remains is to test and demonstrate them on-farm and then farmers will adopt them. In fact, the situation is much more complex than can possibly be revealed by the initial diagnostic work, and outsiders really need to spend a year or more living and working in an area to be competent in designing appropriate solutions. As this is simply not possible for most professional researchers, the importance of involving local people in the ongoing research process is once again underlined".

In the final analysis, it is doubtful whether any amount of improvement in the consultative approach can be a substitute for local participation in the technology research and development process. Even given the best scenario, researchers are unlikely to be able to anticipate every aspect of farmer response to a new technology. Although iterative consultation procedures can be self-corrective, it may be more efficient to avoid problems than to solve them. This can be done by making the whole process more participatory from the start.

Participatory On-farm Research

The essence of on-farm research under the participatory paradigm is the willingness of researchers to give up some of their traditional decision-making autonomy in favour of a more interesting dialogue with their farmer colleagues. Over time, participatory variants of ICRAF's D&D procedures evolved both within and outside ICRAF. One of the chief features of the new participatory methods was the use of group processes to "socialise" the D&D procedures and devolve more of the decision-making to the local communities.

Socialising On-farm Research

The simplest way of making a consultative process more participatory is to reverse the usual proportion of researchers and farmers. In consultative research, a multidisciplinary team of three to seven researchers usually interviews a single farmer. This tends to be overwhelming for most farmers, with the result that they hold back. Reversing the imbalance by sending one or two researchers to participate in a meeting of ten to 20 farmers completely changes the dynamics of the interaction. The farmers discuss more among themselves rather than simply responding to the questions of the researchers and the information that emerges bears a far greater resemblance to what farmers actually think.

Forestry Research within CGIAR

The Consultative Group on International Agricultural Research (CGIAR) is an informal consortium of more than 40 donor agencies, together with representatives of developing countries, elected through the FAO regional agricultural commissions. CGIAR was established in 1971 to support a system of agricultural research around the world.

The early work of the International Agricultural Research Centres (IARCs) associated with CGIAR was largely focused on increasing the productivity of food crops. Given the rapid growth of urban populations in developing countries, two or three times faster than the rate of increase of rural populations, and the associated shift in diet towards grains that can be cooked quickly and stored easily, the IARCs' concentration on productivity was both logical and appealing to governments.

The earlier CGIAR centres are best known for their development and promotion of the "green revolution" crops of rice and wheat, which included varieties designed to be especially responsive to fertiliser and irrigation. CGIAR-developed varieties of staple food crops and their derivatives made it possible for developing countries to produce three to four times as much food as they did in the 1950s.

However, by the mid- 1980s, CGIAR recognised that its main focus needed to be supplemented by a greater attention to natural resource conservation and management as well as the sustainability of agricultural production. The benefits of CGIAR-sponsored research were not reaching a sufficiently large proportion of the intended clients either because the technologies were too expensive or too risky or because they were unsuitable for the marginal conditions under which many millions of farmers have to operate. Moreover, a CGIAR Committee on Sustainable Agriculture which reported in 1987 that it was not enough to resolve problems occurring in the farmers' own fields and herds.

Farmers also depend on reliable supplies of water for farm and domestic use, as well as on security against soil erosion, siltation and flooding, market gluts and unstable prices. Many farmers, especially the poor, need trees in mixed farming systems to improve and stabilise their agricultural production. In many regions, they also require woodlands and forests as a source of products for domestic and farm use, as a source of germplasm of important agricultural and forest crops and as a foundation for off-farm employment and income generation.

Therefore, there was a move towards a greater emphasis on research in the areas of sustainability, resource management and environment. The Committee's work contributed to the preparations for the expansion of CGIAR from 13 to its current 18 centres. International discussions held before, during and after the 1992 United Nations Conference on Environment and Development (UNCED) further highlighted the interrelationships between poverty alleviation and the use and conservation of forests in the tropics.

These discussions focused the needs of forestry research clearly into the heartland of the CGIAR system. The prime stakeholders for forests were the same poor people whose food needs the CGIAR system had been established to meet. The integration of forestry into production systems and the question of sustainability became central to the debate.

In July 1987 in Bellagio, Italy, an international strategy meeting on tropical forests was convened under the auspices of the Rockefeller Foundation, FAO, the World Bank, the UNDP and the World Resources Institute (WRI) to discuss global action to address topical deforestation in the ambit of the Tropical Forests Action Programme (TFAP).

The purpose of the meeting was to debate the constraints to effective implementation of the TFAP (particularly at the national level) and recommend steps to overcome them. One key point to emerge from "Bellagio I" was that the weakness of existing tropical forest-related research was a major hindrance to TFAP implementation and, therefore, to the achievement of sustainable use of tropical forest resources.

In line with this point, in early 1988 the Rockefeller Foundation, the World Bank, the UNDP and FAO jointly sponsored an International Task Force on Forestry Research

(ITFFR) to review research priorities and consider options for strengthening institutional support. The priorities identified by ITFFR were:

- forestry's role in agroforestry, watershed and arid zone land-use management;
- natural resource conservation and management;
- tree breeding and tree improvement;
- utilisation and market research;
- policy and socio-economic research.

In terms of institutional options, the ITFFR report set out several options, including: creating an independent world centre for the direction, execution and coordination of tropical forestry research; expanding the mandate of the CGIAR to include forestry research; and establishing a new consultative group or similar body with a specific mandate for forestry research.

The report was discussed at a second international forestry meeting, held at Wiston House in the United Kingdom in late November 1988. The participants at Bellagio II endorsed the ITFFR recommendations on research priorities and, after much discussion, recommended that forestry research be incorporated into an expanded CGIAR system. Two panels, on forestry and on agroforestry, were established by CGIAR's Technical Advisory Committee (TAC) in early 1989 to conceptualise the research agendas and examine possible institutional arrangements for the incorporation of forestry into the system.

After intensive consultation with national institutes and leading forestry scientists throughout the world, a decision was taken to invite the International Council for Research in Agroforestry (ICRAF) to join CGIAR with an expanded mandate as a global institution for strategic agroforestry research. ICRAF was admitted to the system in 1991 and changed its name to the International Center for Research in Agroforestry.

Also in 1991, the decision was taken to create a new CGIAR centre, the Centre for International Forestry Research (CIFOR), with a global mandate for strategic and applied research on forestry and forest systems, and lead responsibility for coordination of forestry research within the CGIAR system.

Establishment of CIFOR

The Australian Center for International Agricultural Research (ACIAR) was commissioned by CGIAR in May 1991 to do the preparatory work for the establishment of the forestry centre, a task accomplished in just under two years. In March and April

1993, the Governments of Australia, Sweden, Switzerland and the United States signed an agreement sponsoring the legal establishment of CIFOR while 17 international donors agreed to provide financial support. A Host Country Agreement was negotiated with the Government of Indonesia, which is providing temporary headquarters in its Forest Research Institute at Bogor until a new permanent headquarters for CIFOR's international activities can be constructed, also in Bogor.

As defined by its constitution, CIFOR's mission is to "promote the sustained wellbeing of people in developing countries, particularly in the tropics, through collaborative strategic and applied research in forest systems and forestry, and by promoting the adoption of improved technologies and management practices". A key element of the centre's mandate relates to increasing the forestry research capacity of developing countries.

CIFOR has gone through an intensive process of consultation with forestry researchers throughout the world in order to prepare a strategy and a medium-term plan for research from 1994 to 1998. Twenty-five national and regional seminars were held and more than 150 individual scientists around the world contributed to preparation of the plan. CIFOR will concentrate its research and related activities in five programmes and, in addition, there will be a strategic research planning component located in the office of the director-general.

The consensus to emerge from these consultations was that the greatest payoff and impact would be likely to come through policy analysis and development. CIFOR would conduct surveys and experiments to provide the biophysical data to underpin policy development while the needs of policy research would provide the demand pull for the centre's biological and technological research programmes.

Throughout the period of CIFOR's establishment, there has been a general recognition that it must operate in a highly decentralised manner. The centre's research will need to produce results that can be generalised to apply to a wide range of tree species, biophysically defined sites, management objectives and socio-economic conditions. It would clearly be impossible to carry out the wide range of research implicit in CIFOR's mandate from a single headquarters location.

From the outset, CIFOR has been conceived as a "centre without walls". It will operate in close collaboration with national forestry research systems in the developing countries and with relevant specialised institutes in the industrialised world. Very little research will actually be carried out at CIFOR's international headquarters in Indonesia; it is anticipated that 70 percent of its resources will be deployed away from the Indonesian office.

CIFOR's headquarters staff will travel frequently to work with their partners in the rest of the tropical world and these partners will also have the opportunity to spend time at CIFOR's headquarters to work with the multicultural team of scientists located there. It is expected that most of the research programmes will include partners in both the developed and developing countries, thus the basic operational model will be a tripartite series of activities. Several such programmes are already under way; for example:

- work on forest policy in India, particularly focusing on joint forest management, is being developed in collaboration with the University of Florida in Gainesville, the Indian Forest Service and the Tata Energy Research Institute in New Delhi;
- case-studies on the long-term sustainability of forestry-related rural development projects in the Amazon basin are being developed by the WRI in Washington, DC, supported by the United States Forest Service and BMZ (Germany) and in collaboration with Empresa Brasileira de Pesquisa Agropecuária (EMBRAPA) as well as a variety of non-governmental organisations in the region;
- ACIAR is collaborating with forest research institutes in Malaysia, Indonesia, the Philippines and Thailand in a CIFOR project aimed at examining the problems inherent in the reforestation of *Imperata* grasslands in these four countries;
- CIFOR is working with FINNIDA, the World Bank and forest research institutes in Zimbabwe, Malawi and other Southern African Development Community (SADC) countries to examine the problems of community management of dry woodlands in southern and East Africa.

CIFOR's success or failure will be largely a function of the impact on its developing country partners. One of its first major research activities will involve the International Service for National Agricultural Research (ISNAR) in the Netherlands for carrying out a baseline study of forestry research capacity and effectiveness in the tropics. Methodologies for evaluating both CIFOR's impact on research capacity and the impact of research on the welfare of the ultimate beneficiaries will be developed at ISNAR's headquarters in The Hague.

ICRAF's Role and Future Research Agenda

Agroforestry is probably the most complicated biological challenge for CGIAR, i.e. how to grow annual crops with trees in such a way that the inevitable competition for light, water and nutrients results in sustainable food production without degrading the environment. Agroforestry farming systems are among the more attractive sustainable alternatives to shifting cultivation. They are also critical for the reclamation of secondary forest fallows and derived grasslands which often follow in the wake of short-rotation shifting cultivation.

At the farm level, on-farm tree planting and improved on-farm tree management can play a major role in increasing farm productivity, raising farm incomes, improving food security and conserving farm, soil and water resources. There is a limited understanding of how such interactions occur at the process level and how they can best be manipulated by adapted germplasm and agronomic and silvicultural management techniques.

Furthermore, agroforestry systems need time to get established and to exercise key functions such as nutrient cycling and soil conservation. Much of ICRAF's work at priority locations in humid, semi-humid and semi-arid ecosystems is aimed at developing a predictive understanding of the major interactive processes between people, trees, crops and/or animals that will be available to or of value to each agro-ecosystem.

ICRAF has given special emphasis to socio-economic methods for studying local peoples' perception of agroforestry systems. It has developed a strong agroforestry network system with national research institutes in Africa - the Agroforestry Research Networks for Africa (AFRENAs) - and has plans to develop similar networks in Asia and Latin America.

Major gaps exist between traditional agroforestry practices and improved agroforestry technologies. ICRAF's overall strategy is aimed at overcoming these gaps through four research and three dissemination programmes. Research programmes will be implemented using three operational modalities: activities at headquarters, collaborative networks and eco-regional mechanisms.

In Africa, the AFRENAs will be consolidated; in Latin America and Asia, ICRAF will link into existing institutional structures dealing with agroforestry and, at least initially, will focus on the humid tropics. Appropriate laboratory, greenhouse and field research facilities will be established in Nairobi to enable ICRAF to fulfil its global strategic research and dissemination mandate.

Partly in response to the increasing concern about sustainable agriculture and partly in response to the perceived need to take a more ecoregional approach to re search, several IARCs have forestry and/or agroforestry research programmes in progress. Some of the research is specific to individual IARCs but there is an increasing tendency for research on renewable natural resources to be undertaken by several IARCs working together with national agricultural and forestry research organisations.

A notable example is the global project "Alternatives to Slash-and-Burn Agriculture" (ASB), coordinated by ICRAF and involving the International Centre for Tropical Agriculture (CIAT), CIFOR, the International Food Policy Research Institute (IFPRI) and the International Institute of Tropical Agriculture (IITA), together with national institutes in Africa, Asia and Latin America. Throughout the dialogue on how the CGIAR system can most usefully contribute to global forestry research needs, a conscious effort has been made to maintain or to introduce forestry and agroforestry research components into the agendas of the centres.

Agenda 21, one of the outputs of UNCED, has also stimulated the IARCs to take a broader view of their potential roles, and the current medium-term plans usually indicate which specific areas of Agenda 21 might be assisted by an individual IARC. Brief summaries of the forestry- and agroforestry-related research of the other CGIAR centres follow.

CIAT

The International Center for Tropical Agriculture in Colombia was the first IARC to move strongly towards an ecoregional approach in its research planning. Its Forest Margins programme will undertake research on both sides of the forest-agriculture boundary but anticipates close collaboration with CIFOR on the forest side. ICRAF and IFPRI are likewise partners with CIAT at ASB project sites in the western Amazon.

CIP

The International Potato Centre (CIP) in Peru is about to launch an ecoregional programme in the high Andes and may become involved with agroforestry, if not with forestry. CIP is in contact with the FAO-supported regional forestry research project, INFORANDES, in Ecuador.

IBPGR

The International Board on Plant Genetic Resources (IBPGR) in Italy has undertaken germplasm collections and characterisations of fruit trees for several years. Recently, the IBPGR has sought to capitalise on its experience in ex *situ* conservation and seed management by incorporating forest trees into its ambit. Meetings to clarify the respective roles of CIFOR, the IBPGR, ICRAF and FAO were held in Rome in 1991, Nairobi in 1992 and, most recently, Rome again in July 1993. The IBPGR has appointed a forestry germplasm specialist to its staff and has been involved in four major consultancy studies intended to clarify the issues and suggest future directions for research.

ICRISAT

The International Crops Research Institute for the Semi-Arid Tropics (ICRISAT) in India has undertaken some notable ecophysiological work on alley cropping at its headquarters in Hyderabad. At its Sahelian centre in Niamey, the Niger, work has been undertaken on *Faidherbia albida* agroforestry systems, and on shelterbelts in the Sahelian countries.

IFPRI

Located in Washington, DC, IFPRI has sponsored work on the labour requirements of fuelwood collection and is now involved in many aspects of policy research on renewable natural resources and environmental matters at both micro and macro levels. It has cosponsored and been a principal organiser of four workshops on research agendas for tropical land-use and forestry policies and is collaborating on policy research with ICRAF on the ASB, especially in the western Amazon basin. IFPRI's new Environment and Production Technology Division is undertaking research on arresting deforestation and resource degradation in forest margins of the humid tropics and is examining both technology and policy options.

IITA

The IITA in Nigeria is well known for its long-term studies of alley cropping. It is a principal collaborator with ICRAF on the ASB project in West Africa, using its humid tropics substation at M'balmayo in Cameroon particularly to investigate the components and interactions within agroforestry systems.

ILCA

The International Livestock Centre for Africa (ILCA), headquartered in Ethiopia, has made collections and tests of forage shrubs and trees for many years. ILCA collaborated with the IITA in Nigeria to develop and run a research network on the incorporation of the nitrogen-fixing *Gliricidia septum* into agroforestry systems, both cropping and forage.

The network had success especially as a result of its good leadership, the timely delivery of resources to collaborators and the application of lessons from other agricultural and forestry research networks. It may well provide a model for future forestry networks within the CGIAR system.

IRRI

The International Rice Research Institute (IRRI) in the Philippines has provided assistance to other institutions in the Philippines for research on mixed cropping systems, especially in the development and adaptation of Sloping Agricultural Land Technology (SALT).

ISNAR

The mandate of the International Service for National Agricultural Research (ISNAR), located in the Netherlands, encompasses national forestry research systems alongside those in agriculture, drainage and irrigation and fisheries. ISNAR has included national forestry research systems in its regional programme in SADC countries to reorganise the institutional basis for research. ISNAR's compilation of agricultural research indicators, its development of a project-based system for planning, control and accounting of research as well as its work on evaluation and assessment of impact are all highly relevant to forestry research organisations. Organisations seeking to strengthen national forestry research services can learn much from ISNAR's experiences in almost 50 developing countries.

WARDA

The West Africa Rice Development Association (WARDA) in Côte d'Ivoire has expressed interest in working with CIFOR on the hydrological aspects of managing upland watersheds that feed the rivers supporting inland swamp rice culture.

Collaboration with National Research Systems

Because of the high degree of interdependence between CGIAR strategic work in forestry and agroforestry and adaptive research at the national level, much thought has been given to the development of improved mechanisms for supporting effective networks with national forestry and agroforestry research institutions.

An initiative of FAO, the Asian Development Bank and the UNDP, the regionally funded and managed Forest Research Support Programme for Asia and the Pacific (FORSPA), offers promise for replication elsewhere. FORSPA provides a formal mechanism for linking up and sharing research interests of national forestry research organisations. In so doing it could also become an effective research communication and consultation network, with which CGIAR and its centres could collaborate at a relatively low cost compared with the effort of setting up one-to-one linkages with every national forestry research system.

Ways and means of strengthening similar collaborative linkages with national forestry and agroforestry institutions in other regions are currently being explored. The possibility of initiating a FORSPA-type project in West Africa, building on the experience of the regional tree improvement and seed distribution network, is under examination.

CGIAR Response to Agenda 21

The basis of CGIAR's response to the challenge of UNCED's Agenda 21 is the effort to balance productivity and natural resource management in all CGIAR-supported research. A key element of this response is an ecoregional approach that aims at performing research in and for regionally defined agro-ecological zones.

To deal more effectively with natural resource conservation and management issues as part of agricultural, forestry and fisheries development, CGIAR has brought in new centres and is making significant changes to its structure and mode of operation. These changes are mainly based on:

- the need to strengthen CGIAR's scientific capability in the area of soil, water, forestry and fisheries resource management:
- the recognition that natural resource management issues are often agroecologically site-specific and require more intensive multidisciplinary research that focuses on high-priority regions and has a special emphasis on alleviating rural poverty, all undertaken as part of ecoregional research;
- the importance of interacting with local people and, particularly in marginal ecosystems, building research on traditional knowledge, requires that the CGIAR centres further strengthen their capability of tackling socio-economic and macroeconomic policy research;
- the recognition of the role that women play in decision-making at the farm and household level has urged CGIAR to focus more on gender issues;
- the need to help national research institutes strengthen scientific expertise in natural resource management.

Above all, CGIAR has launched a process that will determine how a substantive environmental capacity can be integrated into the programmes of all its centres. These major efforts can enhance CGIAR's participation in the post-UNCED process. Their success, however, depends not only on CGIAR but on the continuing commitment of the international community to attaining UNCED's goals.

REFERENCES

Ashby, J.A., 1990. *Evaluating technology with farmers: A handbook.* CIAT Publication no. 187. Obtain from CIAT, Apartado Aereo 6713, Cali, Colombia. 95 pp. Available from ciat@cgiar.org

Deb Roy, R. & Pathak, P.S. 1983. Silvipastoral research and development in India. *Indian Rev. Life Sci.*, 3: 247-264.

FAO 2002 *Joint action learning between farm foresters, private industry, NGO research and extension agencies. Proceedings of a CIFOR/FAO workshop.* Bogor May 2002 (in press)

Nair, P.K.R. (1979). Agro forestry Research: A retrospective and prospective appraisal Proc. Int. Conf. International Cooperation in Agro forestry. ICRAF Nairobi, pp. 275-296.

Shah, P, 1988, Potential of Agroforestry as land use technology for rural poor. Advances in Forest Research in India, 1:27-100 p.

Heinrich, G.M. 1992. *Strengthening farmer participation through groups: Experiences and lessons from Botswana. On-Farm Client-Oriented Research* Discussion Paper No. 3. ISNAR. 31 pp

Van Crowder, L and Anderson J. 1996. *Integrating Agricultural Research, Education, and Extension in Developing Countries*

Van Crowder, L 1996. *Decentralised extension: effects and opportunities.* SDdimensions FAO

Bibliography

Anderson,J. 1999 "Pluralism and sustainable forestry and rural development". Proceedings of an international workshop 9-12 December . FAO 1997.

AnilKumar, A., Babu.B and Ramachandran,V., 1999, "Attitude of farmers towards agroforestry programmes in Kerala". *Indian Journal of Forestry,* 22(2):155-159 p.

Arnold, J.E.M. 1990. "Tree components in farming systems". *Unasylva* 41(160).

Ashby, J.A., 1990. *Evaluating technology with farmers: A handbook.* CIAT Publication no. 187. Obtain from CIAT, Apartado Aereo 6713, Cali, Colombia. 95 pp. Available from ciat@cgiar.org

Austria, I. 1995. "Developing a community-based marketing information system for integrated forestry products: the Philippine experience". *Unasylva* 48(183).

Curtis, A and Race D, *1998 Links between Farm Forestry Growers and the Wood processing Industry. Lessons from the Green Triangle, Tasmania and Western Australia,* RIRDC publications no 98 /41

David, Soniia 1995. "What do farmers think? Farmer evaluations of hedgerow intercropping under semi-arid conditions". *Agroforestry Systems* 32: 15-28.

Deb Roy, R. & Pathak, P.S. 1983. "Silvipastoral research and development in India". *Indian Rev. Life Sci.,* 3: 247-264.

Denholm, J. 1991. *Agroforestry in the mountain areas of the Hindu Kush-Himalayan region.* ICIMOD Occasional Paper No. 17. International Centre for Integrated Mountain Development, Kathmandu.

Dwivedi, A.P. and Sharma, K.K. 1989. *Agro forestry: its potential.* Paper in Seminar on Eucalyptus, FRI, Dehradun.

Eastin, I. 1996. *The role of marketing in promoting a sustainable trade in forest products in tropical regions.* Paper presented at the FAO/UPLB Workshop on Forest Products Marketing, Los Banos, Laguna

FAO.. 2002 (1) *Equitable partnerships, markets and forest extension.* Forestry Information Note.

Franzel, S. 1999. "Socioeconomic factors affecting the adoption potential of improved tree fallows in Africa". *Agroforestry Systems Vol 47, 305-321.*

Hagmann, J., Chuma,E., Murwira,K., Connolly,M. 1999 *Putting Process into Practice: Operationalising Participatory Extension* AgREN Network Paper no 94 ODI July 1999

Heinrich, G.M. 1992. *Strengthening farmer participation through groups: Experiences and lessons from Botswana. On-Farm Client-Oriented Research* Discussion Paper No. 3. ISNAR. 31 pp

Jahnke, H.E. 1982. *Livestock production systems and livestock development in tropical Africa.* Kiel, Germany, Kieler Wissenschaftsverlag Vauk.

Kane, B.T., Wilson, G.F. & Lawson, T.L. 1984. *Alley cropping: a stable alternative to shifting tiny cultivation.* Ibadan, Nigeria, IITA. 22 pp.

King, Christine A. 2000 *Systematic processes for facilitating social learning. Challenging the legacy.* Agraria 233 Swedish University of Agricultural Sciences

Mann, H.S. & Saxena, S.K., eds. 1980. *Khejri* (Prosopis cineraria) *in the Indian desert its role in agroforestry.* CAZRI Monograph No. 11. Jodhpur, India, CAZRI. 93 pp.

Mayadill, H.T., 1991, Agroforestry for Tropical Rainforests. Institute for Tropical Rain forest and Ecology. Agroforestry systems. 13(3):259-267 p.

Nair, P.K.R. 1979. "Agro forestry Research: A retrospective and prospective appraisal". Proc. Int. Conf. International Cooperation in Agro forestry. ICRAF Nairobi, pp. 275-296.

Olofson, H. 1983, Indigenous *Agroforestry Systems. Philippine Quarterly of Culture and Society,* II: 149-174.

Schirmer, A. 1983, *The role of agroforestry in the Pacific. Suva, Fiji, Univ. South Pacific.*

Shah, P, 1988, "Potential of Agroforestry as land use technology for rural poor". *Advances in Forest Research in India,* 1:27-100 p.

Sharma, K.K, 1996, "Agroforestry in farming systems development". *The Indian forester,* 122 (7): 547-599 p.

Swindale, L.D. 1982. Distribution and use of arable soils in the semi-arid tropics. In *Managing Soils Resources - Plenary Session Papers.* Proc. 12th Int. Congress Soil Sci., 816 February 1982, New Delhi.

Tewari, D.N. 1995. *Agro forestry for increased, productivity, sustainability and poverty alleviation.* International Book Distributors, Dehradun.

Van Crowder, L 1996. *Decentralised extension: effects and opportunities.* SDdimensions FAO

Van Crowder, L and Anderson J. 1996. *Integrating Agricultural Research, Education, and Extension in Developing Countries.*

Index